WORLDS at STAKE

WORLDS at STAKE

Climate Politics, Ideology, and Justice

AARON SAAD

Fernwood Publishing
Winnipeg & Halifax

Development editing: Fiona Jeffries
Copyediting: Erin Seatter
Cover design: John van der Woude
Type setting: Jessica Herdman

Printed and bound in Canada

Published by Fernwood Publishing
2970 Oxford Street, Halifax, Nova Scotia, B3L 2W4
and 748 Broadway Avenue, Winnipeg, Manitoba, R3G 0X3
fernwoodpublishing.ca

Fernwood Publishing Company Limited gratefully acknowledges the financial support of the Government of Canada, the Canada Council for the Arts, the Manitoba Department of Culture, Heritage and Tourism under the Manitoba Publishers Marketing Assistance Program and the Province of Manitoba, through the Book Publishing Tax Credit, for our publishing program. We are pleased to work in partnership with the Province of Nova Scotia to develop and promote our creative industries for the benefit of all Nova Scotians.

Library and Archives Canada Cataloguing in Publication

Title: Worlds at stake : climate politics, ideology, and justice / by Aaron Saad.
Names: Saad, Aaron, author.
Description: Includes bibliographical references and index.
Identifiers: Canadiana (print) 20220259852
Canadiana (ebook) 20220259933 | ISBN 9781773635644 (softcover)
ISBN 9781773635866 (EPUB) | ISBN 9781773635873 (PDF)
Subjects: LCSH: Climatic changes—Political aspects.
LCSH: Climatic changes—Moral and ethical aspects.
LCSH: Environmental justice.
Classification: LCC QC903 .S23 2022 | DDC 363.738/74—dc23

CONTENTS

For Alexandria, always.

ACKNOWLEDGMENTS

I WANT TO START off by thanking Fiona Jeffries, my editor at Fernwood. Writing a first book is a daunting endeavour, but Fiona's feedback, guidance, understanding, and encouragement turned the process into something like an adventure. Thanks also to the rest of the team at Fernwood for all their hard work bringing this together.

I am indebted to the book's early reviewers, David Camfield and Justin Podur, whose careful, insightful, and invaluable suggestions made this work truly and immeasurably better. Shout-out as well to Erin Seatter for her copyediting work spotting more errors and unclear passages than I care to admit had made it into my draft.

I didn't know it at the time, but the idea for *Worlds at Stake* started forming while I was working on my PhD at York University. And so an enormous thanks is owed to the professors I was fortunate enough to work with and who helped expand and refine my thinking about climate politics and justice, in particular Ellie Perkins, Terry Maley, Idil Boran, Anna Zalik, Liette Gilbert, Stefan Kipfer, Kaz Higuchi, and Ilan Kapoor. My brilliant York colleagues Michaela McMahon, Jen Mills, and Kasim Tirmizey also deserve acknowledgement for helping me think through so many of the ideas that made it into this book.

Finally, thanks always to my parents, Abe and Linda, whose constant love, wisdom, and support have made my journey here (and everywhere) possible.

PART 1

....

ESSENTIAL GROUNDWORK

1

A NEW WORLD UPON US

I CAN HEAR CLIMATE change in my sister's cough.

Her lungs tend to disagree, fitfully, with something in the smoke that suffuses the sky over our hometown when the wildfires burn uncontrollably, as they tend to do in these new summers. In those moments, the air over the place we grew up takes on a different aspect as it saturates with the particulate of so many blazing trees; breathing it in inflicts and assaults. It's this sound that has come to mark a cleaving point in my life — the rupture and breach — distinguishing a time when climate change was something I only read and wrote about abstractly from a time when it became viscerally and disturbingly real to me that we are no longer living in the same world.

And what about you? How have you come to know this new world? There are, after all, so many ways now.

Some of them are subtler. One testament to the arrival of a different earth is a changing lexicon. New realities, after all, demand new words — whether created, repurposed, or pulled from obscurity. In the lethally blistering summer of 2021, there came a need for terms like *heat dome* (when high atmospheric pressures lock heat over an area for days over even weeks), *wet-bulb temperature* (the point where ambient air becomes too warm and humid to receive heat from sweating skin) and *pyrocumulonimbus* (when intense wildfires generate their own clouds, themselves capable of throwing fire-starting lightning earthward). That fall, as sections of British Columbia, Canada, were inundated, media referred to supercharged *atmospheric rivers* hanging in the sky like deluging swords of Damocles. The years before introduced us to *climate anxiety* and *climate grief* and even *solastalgia* (Michelin 2020), a homesickness that sets in without ever having to leave home — because the climate that once informed our comforting sense of place has been driven away. And of course, new terms had trickled into our language even before that. For years now, countries have been urged to ration and eliminate greenhouse gas emissions in accordance with *carbon budgets,* and we are aided in making more sustainable personal choices by *carbon pricing* or offers to purchase *carbon offsets.* On bookstore shelves, we find *cli-fi* (climate fiction) novels. Governments declare *climate emergencies* and consider *fossil fuel nonproliferation treaties.* Earth-system scientists argue that human impacts on the environment, including through climate change, are so significant they have ushered in a new geological epoch: the *anthropocene* (Hamilton, 2017). (Political ecologists, concerned that the term lays the blame on humanity too broadly, prefer an alternative to name the economic system that is to them the true culprit: the *capitalocene;* [Moore 2015, 169–92].) Amid the visions we have for our collective future is that it be *net zero,* where the more stubborn emissions still escaping from tailpipes decades from now are balanced and neutralized by an equivalent pulled out of the air. We distinguish tolerable from terrifying tomorrows through adjectives built with temperatures. The existential fears that older generations had of looming nuclear holocausts, today's generations see in a *3-degree* (or warmer) future while the utopian imagination is now directed at securing a *1.5-degree* world, one we are sure prevents us from triggering *tipping points* in the climate system.

A far less subtle testament to this new world is the onslaught of shocking instances of climate devastation that we now hear about in abundance. It has become something of a custom in recent years to open books about climate change on an anecdote from somewhere in the world that conveys our new reality of extremes. I thought of doing the same here, but nothing I considered felt right. And I think the problem is this: All climate disasters now quickly become dated. More and worse always loom. The rarity of disasters in the old world — the randomness with which they once broke through the barely permeable limits of probability — has given way to an always renewing wave of roving, punishing climate events.

If you are reading this book, I suspect that one way or another you already know that this is not, climatically speaking, the same world. You know that the seas are rising. You have heard how, in coming decades, millions may have to decide whether to leave their homes, whether due to the press of waters or the encroachment of deserts or the failure of rain. You, having lived through the hottest years on record, have felt the new unbearability of summer, mild, for all its undispellable swelter, compared to what is coming. You have seen images of the storms bending trees at violent angles and the debris of gale-disintegrated homes and schools and roads and farms. You have seen the red and orange and ash-smoke footage of infernos reducing just-now life-brimming ecosystems to ghostly smoke. You have seen children, fearful of and for their future, having to march.

And you might know even more than that. You know, maybe, about how the worst of these effects has been falling disproportionately on communities that did little to cause this and have the fewest means of withstanding it. You know, too, perhaps, how in this unequal world, gender, race, class, Indigeneity, able-bodiedness, place of birth, and other intersecting dimensions of identity that should not matter for our life chances have come to very much matter.

You know, as well, the Faustian bargain at the heart of all of this: in exchange for the powers stored vastly and densely in fossil fuels, humanity is losing the climate that endured since the end of the last ice age, the only world we know with certainty was capable of supporting agriculture and civilization. This devil's deal has proven difficult to back out of. A full generation after the governments of the world began negotiating responses to climate change, the 2020s began with 83 percent of global primary energy still coming from fossil fuels — 31 percent from oil, 27 percent from coal, and 25 percent from natural gas (BP 2021, 11–12).

You know that, already, enough excess energy has accumulated to warm the *entire planet* by just over 1°C. If the politically miraculous can occur, the world will succeed in preventing warming before it exceeds 1.5°C relative to preindustrial times — a still dangerous level, but one that there is a chance of adapting to and that likely leaves some of the climate system's more devastating tipping points from being reached. At the end of the 2021 global climate negotiations in Glasgow, the United Nations Secretary General told the world that all-important target is on "life support." Put together, the policies that governments were willing to pursue would cause the world to warm by a truly devastating 2.7°C (Climate Action Tracker 2021). The Intergovernmental Panel on Climate Change (2021, 29) tells us that, for a two-thirds likelihood of limiting temperature rise to 1.5°C, no more than 400 gigatonnes of carbon dioxide could be emitted from the start of 2020. All the world has to do is keep current levels of emissions flat and that amount will be gone by around 2030.

Knowing all this, perhaps you feel the weight of this terrible historic moment pressing heavily upon you, and want a chance to think about not only what to do, but also what to do that is right — a chance to think about what kind of world we should fight for in this moment when the world is at stake.

"SYSTEM CHANGE, NOT CLIMATE CHANGE"

That is why this book was written. The climate crisis is sounding an urgent alarm alerting us that something about our society — our very way of life — must change, but the nature of that change is the source of tremendous disagreement. There are vastly different ways to hear the alarm of climate change and what it is calling on us to do. Consider the massive questions behind the simple sign ubiquitous at every climate march: "System change, not climate change." Which system is supposed to change? How should it be changed? What should take its place? Why should we change *that* system instead of another? Who is to change it? Once we appreciate that such questions necessarily invite a complex mix of answers on which reasonable people can disagree, we can also appreciate that the climate crisis is necessarily political and that its politics involve more than simply finding some neutral and objectively "best" solution. At their core, rather, lie struggles to shape the response according to competing and intensely held ideals about our political and economic institutional arrangements, the human relationship with the environment, the nature of progress, the appropriate role of technology, what (if anything) we owe to each other, and more.

That's because people hold a plurality of views on what is right and what is wrong on a great variety of social and political issues, views that condense and cohere into some idea of what society ought to be like — what we refer to, more simply, as *political ideologies.* Ideologies play an immensely powerful role in shaping the different ways people understand the world and seek to answer its challenges, including climate change. They guide us in our search for the source of a given social problem, and in our exploration for solutions that uphold or transform society in ways consistent with our political beliefs. *Is the problem due to a minor issue best solved through small reforms in an otherwise ideal system? Or is the problem due to some fundamental element of a very broken system and yet another sign that an entirely new order is required?*

Ideology helps us to figure out what actions are permissible, unthinkable, or radically necessary.

Depending on one's ideological worldview, the crux of the climate crisis can, as we will see throughout this book, have different explanations: a failure of the market to accurately price fossil fuels; an irrational faith in social and economic change that prevents us from embracing technologies that can engineer the climate itself; the dominance of a political and economic system that is opposed to the use of strong regulation, economic planning, and public investment on the part of a democratic state; an unshakable addiction to economic growth on a finite planet; or the pathologies of the capitalist system itself. For some, ideology even affects how willing they are to believe there is a crisis at all. This is why not every solution to the crisis will seem like a solution to each of us: it may not lead to the kind of world that we most want to live in.

What this means, therefore, is that the world is at stake in more ways than in the *environmental* sense of planetary conditions imperilled by a destabilized climate. The use of the word *world* in this book is meant to evoke something inclusive of but also beyond what is evoked by the often-used synonyms *environment* or *planet*. We often say, following some epochal political event like 9/11 or the COVID-19 pandemic, that "the *world* changed." To speak of the world being at stake in this second, *social* sense is meant to get us to think about how — and how well — our lives would be lived under very different versions of society shaped according to different values and made real by different economic and political institutions.

The alarm being sounded by the climate emergency, the one telling us that something about our current way of life — our current world — must now change, is urgent enough that it has created an inflection point in human political history. It thrusts on those of us living today a choice and struggle about which world we want to create for ourselves and as a model for others as we shape the major changes required to answer the crisis.

If we are to determine which of these worlds to fight for, and which to fight to prevent from coming into being, then we must survey the political landscape created by these different and conflicting responses to the climate crisis. This exploratory approach is one that, first, increases familiarity with the breadth of potential political projects in order to broaden the imagination of the possible and how to make it happen. Nothing limits our politics like feeling that the status quo is somehow natural or can only be subject to marginal change — that the way things are now are the way things need to be. Embarking on this journey of discovery will, I hope, give readers a chance to appreciate the range of options for climate responses beyond those privileged by political and economic elites and mainstream media.

Second, as readers become more familiar with the landscape of climate politics and ideologies shaping it, they will be better able to critically evaluate the outcomes and motivations associated with the climate policies proposed by governments, experts, thinkers, media, and movements. Ideas matter, after all; to the extent they are made real in the world, they have consequences. They usher in one world instead of another. We will be investigating the "deep" political content of these programs for responding to the crisis. What priorities are inherent in them? What arrangements of power do they protect? What liberatory potentials do they suppress or nurture? What relationship with the earth do they assume? What worlds, in other words, would they bring about? Questions of this sort are why the content of this book is informed by a climate justice perspective. There are moral implications to pursuing any given climate response that must be part of any critical inquiry.

Finally, this exploratory approach arises from a belief that we hold a tremendous collective political power to shape our world, and for the better. Think of this power as *potential energy*, much of it still stored, still unlocking. In exploring the breadth of climate responses and in being able to imagine critically what kind of society each would bring about, readers can better find their place in the climate struggle and decide how to contribute to the work that potential energy might do.

PLAN FOR THIS BOOK

Part 1 of this book lays down essential groundwork. The next chapter starts from the notion that how we process the climate crisis is affected by how we think the world currently works and how it ought to look. It therefore seeks to familiarize readers with the concept of ideology being used throughout this work. It highlights several essential features of ideologies: their content, their tendency to organize and define that content so that it is integrated as coherent and noncontradictory systems of thought, their concern with shaping a society's institutional arrangements, their powerful role in identity formation, and the "messiness" to them that can powerfully trip up our political thinking. The chapter concludes with a discussion of the concept of the ideological framework — that is, the way ideologies define the shape and boundaries of responses to political issues.

Chapter 3 covers the concept of climate justice. It's a term with no shortage of definitions and uses, but what is common in every instance is a concern with identifying the various moral issues raised by climate change, and prioritizing solutions to those moral issues in any climate response. We survey this range of moral issues that stem from the climate crisis by considering five questions: *Who should do what? Who will be impacted and why? What is the moral significance of climate impacts? Whose views matter and are heard? What is driving the crisis and preventing responses?* The chapter highlights a crucial principle: climate justice demands that we do not accept the existing social and economic order as self-justifying. On the contrary, if those arrangements are preventing an ambitious climate program while also failing to provide a decent and sustainable existence for all, they are subject to potentially radical change.

Part 2 of this book moves on to the first set of ideological frameworks, what we will call the *system-preserving* frameworks due to their underlying concern to respond to climate change without much alteration to the status quo; existing social relations are seen as unproblematic or even approximating an ideal, and so the climate response should uphold as much as possible the way things are. Chapter 4 looks at the neoliberal framework and its suite of primarily market-oriented solutions. Chapter 5 takes on climate change denial, which has dominated the right-wing response. Chapter 6 explores the geoengineering response — the turn to direct climate-intervention technologies — and investigates what ideological currents are shaping those efforts.

Part 3 examines the *system-changing* frameworks. What they share is an analysis rooting the climate crisis in some element of our contemporary way of life that has for too long gone unchanged despite undermining prospects for a decent society. Chapter 7 looks at the social democratic framework, which locates the primary obstacle to climate action in neoliberal hegemony, and counters it with a justice-based Green New Deal. Chapter 8 considers the degrowth framework, which is unique in underlining the role of perpetual economic growth in driving the ecological crisis of which climate change is but one manifestation. It urges us to ponder whether there can be a richer human life without seeking ever more wealth. Chapter 9 features our final framework, ecosocialism, which contributes a series of powerful critiques of existing capitalism and refines and sharpens our critical eye.

Part 4 looks at how we can change the coming world. Its penultimate chapter focuses on the climate movement, surveying a number of its most prominent recent tactics and theorizing their respective contributions. The conclusion offers final thoughts on where the reader might find themselves in the struggle for the climate.

2

IDEOLOGY

S<small>EPTEMBER</small> 21, 2014. N<small>EW</small> York City.
Halfway along the route of what would be declared the largest climate march in history, a reporter asked me why Canadians, like myself, had decided to take part in the People's Climate March. Somewhere between 300,000 and 400,000 people had arrived from across the continent and beyond to make demands of the politicians about to gather at special United Nations talks intended to advance negotiations for a desperately needed global climate agreement — negotiations that had stalled.

I struggled to think of much of an answer. Why *were* we there? Unquestionably, we all wanted something done about climate change. But whether we all wanted the same thing was another matter. On the bus there from Toronto, I had overheard students in York University's Environmental Studies program situating the problem of climate change in capitalism. Perhaps I could tell the reporter that we had come to take on the capitalist system?

But the group that had organized the buses, the Toronto chapter of global climate action organization 350.org, had distributed signs to marchers demanding something less radical and more pragmatic: "Canadians for a Fossil Free World" and "Canadians for Green Energy Investment." Our marching chants, meanwhile, called out Canada's Stephen Harper government ("Hey hey! Ho ho! Stephen Harper's got to go!"), which at the time was aggressively promoting the development of the country's massive oil reserves in the tar sands. Should I tell the reporter Canadians had come to denounce our government in front of the world while adding our voices to those calling for full decarbonization of the economy?

Just prior to the start of the march, I had heard two men somewhere behind me calmly discussing and considering anti-civ philosopher

Derrick Jensen's strategy of armed resistance against not just the fossil fuel industry but civilization itself. That seemed a bit fringe of an answer for me to give the reporter.

And I didn't know it that day, but the organizers had arranged the march in segments to tell a story of the world that could be, to spell out demands for the justice-based economic mobilizations needed to save a habitable climate — demands that would soon find political expression in the Leap Manifesto in Canada and eventually the Green New Deal in the United States.

At the same time, I was aware that not all Canadians held much sympathy for those of us who had come — particularly in Alberta, my province of birth, home to the tar sands and to the highest degree of climate change denial in Canada. How might people who knew me back home react to whatever I answered?

In the end, I don't think I managed to answer the question with anything quotable. But then, the question was a difficult one to answer. No large group of people will want the same things done about climate change because there are just so many different ways we see the crisis based on our political beliefs. It's why we need to get to know ideology.

GETTING TO KNOW IDEOLOGY

Consider the following scenarios:

- Activists have taken it upon themselves to tear down the statue of a prominent political figure who was integral to the founding of the nation but who instituted or upheld state policies that would no longer be considered ethical or acceptable.

- A public school board has decided it will stop teaching about LGBTQ+ issues in its sex-ed curriculum, arguing these are offensive to traditional moral values and inappropriate topics to teach children.

- The ruling government party has passed legislation that will make it harder for specific segments of the population to vote.

- A government decides to put strong restrictions on access to abortion services.

- The owners of a bakery have declared that they will not produce a wedding cake for a same-sex couple because nonheterosexual marriage is against their moral values.

- A set of reliable studies has found that the richest 1 percent of humanity holds around 40 percent of the world's entire wealth.

- As part of dealing with a serious pandemic, a government institutes mask and vaccination mandates.

- A government has instituted a rising price on carbon emissions, making fossil fuels like gasoline for cars and natural gas and coal for electricity more expensive.

It's unlikely that you read these scenarios and felt indifferent towards them. On the contrary, you probably felt that what is going on in each of them was either right or wrong, either morally defensible or cause for concern. And that reaction probably happened not just quickly, but before you could fully articulate why you felt how you did. But with a bit of time, you can start to provide some arguments supporting your initial evaluation of the scenario and offer some thoughts about whether and how to respond to these matters. With a bit more time, you can test those arguments against some of your other beliefs or apply those arguments to different versions of these same scenarios. This phenomenon — encountering a novel situation, experiencing some initial reactions to it, testing those reactions against subtle changes in the scenario, and seeking solutions consistent with your beliefs about what a society ought to be like — is ideology at work.

A central concern of this book is to highlight the major role that ideology plays in climate politics. There are multiple and competing ways of understanding the climate crisis and responding to it. Like the scenarios above, the explanations for why we have failed (and continue to fail) to reduce greenhouse gas emissions, the significance of the impacts of climate change, and the proposed responses are all perceived differently, strongly influenced by ideological belief.

To illustrate, we can imagine responses that would "solve" the climate crisis very rapidly but require such uncommonly extreme ideological beliefs that few people would find them appealing. A first example is to wipe out the human population entirely by introducing, say, some extremely lethal and virulent pathogen. In short order, all human-caused carbon emissions would cease and the rest of life on earth could thrive. Such a response would appeal, however, only to those holding an ideology marked by an extreme nihilistic misanthropy incapable of finding *any* inherent value in our species and by an extreme valuation of non-human life. If that scenario is too extreme, we might seek, alternatively, to cause the immediate collapse of industrial society, which could bring to a close the environmental catastrophes it drives without wiping out the human species, at least not in its entirety. But we should expect little support outside of the few people subscribing to the ideology sometimes called *anarcho-primitivism* or *anti-civ* (e.g., Jensen 2006a, 2006b), which believes human life is best and most ethically lived in preindustrial (and possibly pre-agrarian) societies. A final example of an extreme response is to work towards imposing a totalitarian government that forces people to work at building the postcarbon world in press gangs, severely punishes fossil fuel use outside of a tight quota, and spies on its citizens to ensure no unauthorized carbon emissions occur. But to support such a response would require an ideology that embraces authoritarianism and dispenses with individual freedoms.

The point is this: It is not enough that a response "solves" the climate crisis. It should, in the process, preserve or create a desirable world, and that vision of a desirable world comes from our ideologies.

But before continuing on, a quick side note is required because (as so often occurs when discussing politics) there are competing definitions of key terms. Some readers might be familiar with a different use of the word *ideology* than is used in this book. In some Marxist schools of thought, ideology refers to a tool a society's ruling class uses to control the people being economically oppressed and exploited. That underclass of people has to be made to believe, falsely, that the very same social order oppressing them is actually benefiting them. Otherwise, they might be tempted to overthrow the existing social order. If it's helpful, readers might think of that Marxist sense as "ideology as system of false consciousness imposed by the rulers," and the sense used in this book as "ideology as political worldview." The latter is likely the more familiar sense for most readers, who will have heard of ideologies such as socialism, liberalism, conservatism, libertarianism, and fascism.

Ideology is something we all have. That is because each of us possesses a system of beliefs and values that guides us in thinking about how the world should be and how it should work — about what kind of society we ought to live in. Even people who describe themselves as apolitical are probably not, if by "apolitical" they mean they have no ideology (it's more likely that nothing about the status quo upsets them all that much). To be without ideology, as we are discussing it here, would suggest that a person is so lacking in values as to be indifferent to whether we live in a world marked by slavery, totalitarianism, genocide, racial apartheid, state collapse, and climate catastrophe.

But just because we all have an ideology does not mean we all understand it or how it works. And so to get a better sense of what we mean by it, let's look at several of its key features: (a) the content or "stuff" that ideologies are composed of; (b) the way that ideologies make up coherent, noncontradictory systems of beliefs; (c) their concern with the institutions required to make them real; (d) the role they play in individual and group identity formation; and (e) the "messiness" of ideology. After we do that, we will find ourselves in a better position to understand its role in climate politics.

The Content of Ideology

What makes up ideologies are beliefs concerning the nature of an ideal human society. Ideologies, in other words, are unique groupings of ideas about how our world works and how it ought to. This means that, at their foundational core, ideologies hold some sense of what it means to be human. Are human beings fundamentally good-natured and social, looking to live in rich communities founded on mutual aid? Or are we, as the philosopher Thomas Hobbes famously alleged, prone to mutual distrust and enmity that embroils us in violence unless some authority looms over us? Of all the drives that can animate us — being self-regarding, entrepreneurial, and competitive; being domineering and establish supremacy; being pluralistic, empathetic and nurturing; being freely inquisitive and creative — which make us our "most" human? These are not idle questions. With a sense of our human nature comes a sense of the *human good* that a society ought to uphold in order to make life worthwhile, enjoyable, orderly, fair, or meaningful. If it turns out that humans are fundamentally driven to, say, consume, it would follow that a good society ought to be constructed along consumerist lines with plenty of producers incentivized to provide a vast array of goods and experiences people would want.

But these concepts — human nature and human good — are rather lofty (and potentially even intimidating). They have no single uncontroversial, canonical definition, and so it's understandable that people may not be able to articulate with full precision and confidence how they understand them. Nevertheless, they do play an important role in our political beliefs; dig deep enough into any political worldview, and you will eventually find some engagement with these concepts. Adherents of an ideology, for instance, would not advocate for a society they suspect human beings would be incapable of living in; some sense of human nature has to bound and undergird that vision of a society. Similarly, it would be strange to find an ideology with a vision for society that is unconcerned with quality of life therein or that suggests competing ideologies offer better prospects for human existence; there has to be some underlying sense of the human good it seeks to promote or protect.

While these concepts at the inner core of ideology can be challenging to grapple with, there is, fortunately, more content to ideology that is easier to grasp. Indeed, it's this additional ideological stuff, rather than questions of human nature and ultimate human good, that people tend to appeal to when making political arguments. Perhaps the most common set of concepts we appeal to are *values*, those very highest ideals we feel that a society must realize. Ideologies will differ in the definition, selection, and prioritization of different values in their ideal societies, like freedom, equality, tradition, progress, prosperity, empathy, openness, order, security, and so forth. A related matter that ideologies disagree over is the expansiveness of the *rights* that a society ought to guarantee and which rights should override which, should they come into conflict.

So, too, will ideologies hold different beliefs about what constitutes legitimate forms of political *authority*, how much control they ought to have over individual and collective life, and under what conditions. Ideologies will differ, as well, in their ideals about the nature of people's political *participation* in a society, including how directly they should affect decision-making (e.g., indirectly through representatives or through more direct measures like citizens assemblies and referenda) and how frequently; in which spheres of social life democratic decision-making will occur (e.g., whether economic decisions are to be shaped through participatory means or left to private actors in the market); and what forms of political expression and pressure are acceptable (e.g., whether protest and direct action are permissible in a society). So, too, are there differences in views about *social change*, like its desirability, who is to drive it, in which direction it <u>is to go</u>, and how realistic it is to believe it can happen. There are also intensely competing beliefs concerning *nature and society* and the extent to which human activity ought to be free to reshape, use, and denude the surrounding environment, the moral concern that should be accorded to nonhuman life and ecosystems, and even to what extent it's possible to think of society and nature as separate.

Ideology as System of Political Belief

But all this content cannot come together in just any random way, which brings us to a second important point: ideologies are complex systems of beliefs about the nature of an ideal society in which those beliefs have been integrated with one another in such a way that they hold, for their adherents, an internal consistency. All of the main beliefs within an ideology have to be able to function together to create a coherent view of how the world should work; after all, obviously contradictory beliefs would make an ideology incoherent and implausible. An ideology would not explicitly maintain, for example, that all humans are of equal worth while also maintaining that some races of humans are superior over others. Similarly, it would be difficult to find an ideology insisting on the primacy of human freedom while also arguing for tyranny. The beliefs have to harmonize, so to speak.

None of this is to say that an ideology's opponents will fail to find in it what they think are obvious contradictions. For instance, a socialist might point out a contradiction in a liberal's advocacy of equality under capitalism, the socialist insisting that the two are necessarily incompatible. But this is more of a problem concerning the definition of a key concept than of a fatal flaw in logic — a difference over whether, in this instance, equality ought to mean equality of outcome (guaranteed by rights to income, housing, work, etc.) or whether it ought to be restricted to equality of opportunity (guaranteed by antidiscrimination laws in hiring practice, equal access to primary and secondary education, etc.) complemented by equality in political and social rights (rights to vote, to free expression, to free assembly, etc.).

This is an example of another way that ideologies remain coherent as systems of belief: they "decontest" their key ideas (Freeden 2003, 52–55).

That is, an ideology's followers will by and large agree on how they understand its core concepts, even if outsiders do not. This is an important point to bear in mind as we proceed in this book and describe the core beliefs of different ideologies. Most contemporary ideologies will, for instance, hold freedom as an essential value of an ideal society, but will comprehend in very different ways what exactly freedom means. The same goes for equality (as we just saw), progress, and so on. Standard dictionary definitions will not be all that helpful.

These systems of belief, and the need to maintain their coherency, have powerful effects on our thinking. Through them we interpret novel political events (like climate change). Our ideologies alert us to whether these events pose problems or opportunities with respect to realizing our ideal societies. They guide us in thinking about solutions to a given political problem in ways that are consistent with (and avoid undermining) our visions of how the world should be. Solving a political problem, then, is very unlike, say, solving a math equation, where there is a single, clear, right answer demonstrable to everyone. Different systems for seeing the world lead to different solutions.

To see this at work, let's briefly explore that scenario of the bakery refusing to make a cake for a same-sex couple, and the political problem of whether privately owned companies ought to be free to discriminate in their services based on clients' identities. If you feel that privately owned bakeries *should* have the freedom to deny service to that couple, then should they also have the freedom to discriminate against interfaith or interracial couples? If you feel they should *not* have this freedom, should they also not be free to refuse service to a racist couple that wants a Confederate or Nazi flag on their cake? Must they agree to provide a wedding cake for a polygynist, a zoophile, or a much older man marrying a child bride in, say, parts of the United States where that remains legal?

As you consider questions like these, you begin to think about solutions that are compatible with your system of beliefs. Since no system of belief can maintain both that companies should be fully free to discriminate *and* that same-sex (or interfaith or interracial) couples should be fully free from discrimination, there has to be a trade-off between them. One solution might be to let private companies refuse service as they please; that way, we can live in a society where no one is forced to do anything with their private property that goes against their conscience. But it's not then possible to insist that government ensures everyone lives free of discrimination; anyone discriminated against will simply have to look for other cake options, hoping that the ire directed towards them does not extend society-wide. This solution would be compatible with a system of beliefs that holds extremely high regard for individual liberty, but low regard for government regulation and equality.

Meanwhile, people holding other systems of belief would find a society resulting from the broader application of that solution less than ideal. Another quite different solution might be to make it illegal to discriminate based on what most people today consider morally irrelevant features of identity (sexual orientation, race, religion, etc.), while leaving discretion about the rest up to the bakeries' owners. That way, we can live in a society that has reduced inequality by making some forms of discrimination illegal. This solution prioritizes equality, and permits government regulation to promote it, but, to allow that to happen, some individual freedoms in the market must be constrained, meaning, in this case, some bakers will have to make cakes for same-sex couples even if they don't want to. Any bakery owners uncomfortable with these regulations will simply have to ask themselves whether they want to stay in the industry.

The point is this: which of these solutions, if any, you think is better has to do with further questions about the implications those solutions hold for what kind of world you feel we should live in. And what is true for bakeries and discrimination is also true for climate change — just in immensely more complicated ways. As we saw above with those examples of extreme responses, not every "solution" to the climate crisis will seem like a solution to everyone. As we interpret and evaluate climate responses through that complex system of integrated beliefs about how the world ought to work, we reject those most incompatible with our ideological views and favour a response most compatible with the world we want to see.

Institutional Arrangements

The third feature of ideologies is that they have to be made real in the world. Ideologies are not fantasies or fanciful thought experiments; they are beliefs about how the world really ought to be, how it should remain or become. The task of making the world turn out a particular way requires arrangements that lead people to behave in ways consistent with some vision of an ideal society. It requires, in other words, institutions, like markets, elections, parliaments, constitutions, property rights, legal systems, and so forth, which create and enforce the rules and norms beneficial to a society. And these institutions have to work together to form functioning economic and political systems. A system of institutions that promotes free-market capitalism will lead to a much different society than a system of institutions made to promote a society of equitable capitalist growth and a strong welfare state, just as that system will lead to a different society than one for a steady-state postgrowth economy.

The choice of one of these systems over another leads to vastly different worlds because it holds implications for some defining characteristics of a society: the distribution of power, wealth, privilege, and opportunity; how (and which) people acquire resources to meet basic needs, develop their capabilities, and pursue life goals; the accessibility of high-quality educational, health, legal, and other vital services; the extent to which people can meaningfully participate in decisions that affect their communities; the scale at which energy and resources are used and for what purposes; and the degree of control people have over the conditions of their work. One could go on.

Identity Formation

Ideologies can also come to very powerfully inform our sense of identity. They guide us to see what is morally right and wrong with the world and when to take action to defend what is good. That moral compass is an important part of our sense of who we are. Part of the human condition is wanting to have pride in our individual selves, an important factor of which is believing oneself to be a good person, or at the very least in the right. Even the most evil of individual actions and political causes tend to have rationalizations attached to them that appeal (however unconvincingly) to some greater good. Subscribers to a given ideology will find in it prominent central features that coincide with parts of their identities they would most like to affirm — uncompromising competitiveness or compassionate solidarity, self-interest or generosity, realist pragmatism or idealist utopianism, openness to progress or reverence for tradition. In many instances, our ideological views communicate whose side we are on — the privileged or underserved, the preservers or changers, the oppressed or oppressor, "ours" or "theirs." ____

And the contributions of ideology to identity formation are not limited to an individual sense of self; ideology is also a foundation for group identity. A single individual cannot meaningfully hold a unique personal ideology, just as a single individual would not meaningfully possess a unique language. Ideologies are shared systems of political ideas, providing their adherents with a mutually held group view of the world. To be sure, there may be differences about finer details, but with respect to core beliefs, agreement will be very high. It is on the basis of shared ideology that actors can take action in concert. Ideology alerts collections of allied actors to new or ongoing opportunities or threats to values they together believe are essential to the creation or maintenance of a common vision of a good society.

Ideology Is Messy

Finally, the way ideology "lives" in the mind is complex and messy. The points made above are not meant to leave readers with a sense that holding an ideology is the same thing as holding some perfectly worked out belief system that has been immaculately purged of all contradiction and doubt or is immune to unsound appeals and reasoning.

Like emotion, ideology forms a powerful feature of the human psyche, one that can enhance our experience of the world while guiding us in making sense of and living in it. But it can also lead us astray. Allowed to become dogmatic, ideology can warp or obfuscate the way we process the goings-on of the world. Because ideological beliefs form a foundation of individual and group identities, people are often reluctant to admit when their beliefs might be flawed or poorly founded. To do so would be to lose face (after all, we put our reputations on the line when we outwardly express our political views and attempt to defend them, on behalf of ourselves and our allies, against opposing ones), set off a crisis in one's sense of self, or even risk estrangement or rejection from groups formed on the basis of shared political beliefs. The shared nature of our ideologies means that they come with all the benefits and complications of other sources of group identity. Along with providing a sense of community and purpose, and an extended membership with which to celebrate victories (and mourn setbacks), ideology can pressure us to hold views that we believe we are *supposed* to hold in order to continue to be in good standing with our ideological "side." What we think being a good leftist or progressive or liberal or conservative should entail can influence the positions we are willing to openly hold, defend, and rally for when needed. A particularly wrong take on a given issue — or a pattern of wrong takes — and one risks isolation from their ideological team.

In the face of attacks from opponents, an ideology's adherents have a variety of strategies to defend or hold on to their beliefs, not all of them reasonable or intellectually honest. Ideological threat or insecurity can incentivize people to engage in defensive forms of cognition that guard them from confronting the full implications of inconvenient facts. They engage in what is called *motivated reasoning* or *confirmation bias*, closely related terms for an emotionally driven search for — and acceptance of — only the information that reaffirms already existing beliefs, and a dismissal of evidence that goes against them. Ideological biases can express themselves in still other ways, including selective interpretation of foundational documents (constitutions, religious scripture, classic political books, tracts, speeches, etc.), idiosyncratic understandings of broad political concepts (e.g., freedom, equality), faith that certain features of society are expressions of immutable human nature (e.g., hierarchies, selfishness, aggression), overly certain projections of the consequences — whether dire or desirable — of making particular political decisions, loyalty to partisan news media, and charitable interpretations of or apologetics for the actions of favoured political figures. In other words, ideology can make us selective, consciously or not, about what facts and analyses about the world we are willing to accept and recall and who to believe about them. In the scrum for political dominance, people might even deploy mutually contradictory messages or arguments, the potential cognitive dissonance "resolved" as long as each can be used to advance a higher, more important goal of beating back ideological opponents. (We revisit this last point when we look at the incompatible claims that climate change deniers use to protect their ideological project.)

All of this is a reminder that when engaging with the politics of the climate crisis, converting others to one's views is a fraught challenge. It often involves something much deeper than what can be addressed through, say, presenting them with a string of irrefutable facts or a blueprint of change or a shining vision of a future world. Any of these can be difficult to accept if their content is inconsistent with or undermines the ideological views that form part of a person's identity. It is also a reminder that we ought to be humble enough to question our own political assumptions and beliefs, and be vigilant that we are not using unreasonable strategies to maintain those beliefs in the face of contradictory evidence.

THE LEFT–RIGHT SPECTRUM

Because the terms *left-wing* and *right-wing* are used frequently in the chapters to follow, let us spend some time on them here for readers who might not be confident using them. (Those familiar with them can safely skip ahead to the next section on ideological frameworks.) These terms reflect an effort to illustrate the relationship between different ideologies by placing worldviews along a left-to-right axis, or political spectrum.

What is most commonly being measured along that axis is the degree to which equality or hierarchy is valued by an ideology. The deeper we move into the right wing of the spectrum, the less an ideal society values equality and the more it is characterized by a high degree of hierarchy, where some groups or individuals are seen to deservedly hold more wealth, freedom, power, opportunity, agency, or privilege compared to others. What justifies these status differences can vary. For example, a belief common to conservatives and right-wing libertarians (discussed in Chapter 5) is that differences in economic status are justified because they simply reflect individuals' just rewards for hard work, innovation, or smart use of private property under a well-functioning meritocratic capitalism. Another vision of an ideal hierarchical society might be one in which people deserve higher status simply due to features of their identity — like race, nation, religion, gender, neurotype, or sexual orientation — placing them in an ostensibly superior group that has historically occupied upper social echelons. The more an ideology takes on these latter, supremacist beliefs, the deeper it enters into the far right.

Meanwhile, the more that an ideology's ideal society is characterized by equality, the further it falls on the left wing of the political spectrum. The previous section noted how equality can mean different things. It might be limited to equality of political and social rights, but might also be expanded to include equality of opportunity. These forms of equality tend to be found closer to the political centre. The meaning of equality can be expanded still further to involve equality of outcome — that is, equality in how each person's life turns out. What might be called a *threshold* variant of this sense of equality permits inequality only beyond a minimum level of well-being that is guaranteed universally to all by the state. An *absolute* variant would go even further so that each person in a society experiences effectively equal quality of life and status.

Complicating matters with the left–right spectrum is how people also use it to map out more than just equality–hierarchy. A second set of differences in beliefs illustrated by the spectrum concerns the ideal economic system. The further to the right, the more the economic system becomes a purely capitalist one in which governments play little to no part. Towards the centre, we find more mixed economic systems, with the state taking on an enlarged role by implementing regulations, engaging in economic planning, providing public services, and redistributing wealth generated under capitalism to lower-income individuals and communities. On the left, meanwhile, are variants of socialism, in which the means of production are socially owned and used to meet human needs rather than generate profit for private owners.

Finally, there are ideological differences in attitudes about social change. On the left, social change is embraced to the extent that it eliminates unjustifiable inequalities. If these changes involve the overturning of major institutions to remake the contemporary social order in a far-reaching way — for example, replacing capitalism with some alternative in order to address climate change — they tend to be referred to as radical. (This use of *radical* to describe social change is not to be confused with radical in the sense of extreme, as in "radicalized right wing.") If social changes take place within the existing institutional system, they tend to be described as reformist or progressive. Meanwhile, though the political right is often characterized as discouraging or resisting social change in favour of preserving a status quo (putting the *conserve* in *conservative*), it is probably more accurate to say that social change is justified on the right if it works to restore a traditional hierarchical order (or make society in the image of some imagined or idealized past). This form of change is called reactionary because it tends to occur in response (or reaction) to either revolutionary or progressive victories. In sum, present-day radicals would seek to overturn capitalism, reformists would seek to make it fairer, and, should either succeed, reactionaries would seek to return the capitalist class to the top, and shelter their wealth from any would-be government hoping to tax them to fund its welfare state.

As commonly used as it is, the left–right spectrum is hardly perfect for seeing the richness in differences between political ideologies. Positions on issues such as free speech or the legalization of sex work or recreational drug use can be difficult to place along it. Both left-wing libertarian socialists and right-wing libertarians would, theoretically, be in favour of them. Interested readers can look at the Political Compass (political-compass.org), which adds a vertical axis that measures views on the degree to which the state should have control over individuals. Instead of a one-dimensional axis on which to attempt to place the various political ideologies, the Political Compass produces a grid. Ideologies in favour a capitalist society with large inequality under a state with a higher degree of authority over people's lives would be positioned in the top-right quadrant, those in favour of capitalism under a more libertarian state in the bottom right. Authoritarian socialist ideologies fall in the top left and forms of libertarian socialism in the bottom left.

IDEOLOGICAL FRAMEWORKS

To be able to talk about ideology in the above ways is to help us recognize that when we are thinking about climate change responses, we are no longer dealing simply with a search for some objectively "best" approach, but rather with solutions arising from competing, intensely held, identity-forming belief systems of how the world should be. As we will see, climate change responses are frequently evaluated not only (or even mainly) for their capacity to address the crisis, but also for their ideological content and the implications those responses hold for the world they will preserve or usher in.

Throughout this book, we will look at several different ideological frameworks for addressing climate change. Any framework, real or metaphorical, carries out important functions. During the construction of a house or building, a framework helps establish the structure's boundaries, defining the space it will occupy and how far outwards it will go and will not go. The framework also underlies and supports the structure, ensuring it is supported against collapse as it is built. Similarly, an ideological framework shapes the analysis of a given complex social problem to ensure it can be solved in a way that is consistent with an ideology's core beliefs and does not change society in ideologically undesirable ways. In doing so, it defines and limits which political and economic institutional changes are permissible. It establishes what is to be prioritized in that response and what can be left out.

Let's relate this matter back to the climate crisis. If you believe the societies composing the Global North are the ones best capable of promoting the human good, you would likely be uncomfortable with massive changes to them in the name of fighting climate change. To you, a climate response should not radically alter the way the economy works or change the nature of authority or political participation. Such transformations would not fit into your ideological framework. And the reverse is also true. If you believe that the societies of the Global North are far from ideal, you are probably much more open to fundamental changes that would address the climate crisis *and* usher in what, to you, would be a better society. The chapters that follow show how different ideological frameworks for addressing climate change differ in their treatment of the market, economic growth, capitalism, and even climate science.

But this book is not about surveying different ideological frameworks neutrally. Some means of evaluating the role they play in addressing the moral issues raised by climate change is also called for. For this, we turn to the concept of climate justice.

3

CLIMATE JUSTICE

"What do we want?"
"Climate justice!"
"When do we want it?"
"Now!"

If you go to a climate march, you will almost certainly hear that mobilizing chant shouted confidently and defiantly. You will notice, also, that there is no third question: "What do we mean by 'climate justice'?"

Climate justice, like *ideology*, is a complex term that has over the years been used in many ways. In this book, climate justice involves (a) the identification of the moral issues that are either causing, caused by, or otherwise raised by climate change, and their significance, and (b) the search for solutions to those moral issues. Together these two elements act as a lens, giving us a heightened sensitivity to the moral implications of various matters of climate politics.

A FIRST FACET OF CLIMATE JUSTICE: IDENTIFYING MORAL ISSUES

To understand the range of issues that crop up, let's look at five questions of climate justice:

- Who ought to do what?
- Who will be impacted and why?
- What is the moral significance of climate impacts?
- Whose concerns matter?
- What is driving the crisis and preventing responses?

Who Ought to Do What?

Climate change is occurring in a highly unequal world characterized by a subset of advanced capitalist countries with high per capita incomes (the "developed world" or "Global North"), a much larger group of countries with per capita incomes too low to ensure every person can have their basic needs met ("developing countries" or "Global South"), and a set of emerging economies sometimes referred to as BRICS: Brazil, Russia, India, China, and South Africa. (Of course, these are not perfect or uncontroversial terms, and some alternatives that readers might elsewhere encounter are the "overdeveloped" or "minority" world to refer to advanced capitalist countries, and the "majority" world to refer to countries of the Global South.)

Determining Fair Duties: Differential Responsibility and Capacity

The climate reacts the same way to any given greenhouse gas molecule regardless of the reason it was emitted. But this does not mean that all emissions are morally equal. When it comes to determining whose emissions need to be addressed first, the terms *luxury* and *subsistence emissions* are often used to make an important moral distinction between different sources of carbon pollution. Survival emissions are caused by activities of the poor required for meeting basic needs, like methane release from livestock or rice paddies; luxury emissions, in contrast, are those resulting from the excesses of fossil-fuelled wealth, like gas-guzzling automobiles or regular flights to tropical locales. Any just response to climate change would prioritize eliminating the latter sort of emissions, ones found disproportionately among the rich.

There is long-standing and broad agreement that developed countries ought to take on the earliest and deepest obligations for at least three reasons. First, their contributions to cumulative greenhouse gas emissions make them primarily responsible for the climate change currently being experienced. Historical responsibilities for carbon emissions are extremely unequal. According to figures from James Hansen and Makiko Sato (2021), just six developed countries account for 41 percent of cumulative emissions from 1751 to 2020: the United States (24.4), Germany (5.4), the United Kingdom (4.7), Japan (3.9), and Canada and Australia (combining for 3). The entirety of Africa accounts for just 2.9 percent and India 3.4 percent. (Similar figures are found in Ritchie

2019.) But the role of emerging economies complicates matters. China is now the world's top annual emitter (and accounts for 13.8 percent of historical emissions) but with much of those emissions in service of production for consumption in the Global North.

An interesting study by Hickel (2020c) took a slightly different approach. Rather than look at each country's contribution to total historical emissions, Hickel determined how much each country was responsible for the current climate breakdown. To do so, he assessed how much a country's emissions exceeded or fell below what its fair share would be in a hypothetical scenario in which the countries of the world had divided a safe level of total emissions according to population. The Global North was responsible for 92 percent of the breakdown.

Second, developed countries' emissions have taken up so much of the global atmospheric space that they left others with little to use for their own economic development powered by fossil fuels. Those two reasons suggest that industrialized countries owe what is frequently called a "climate debt" to developing countries (Simms 2009; Klein 2014a, 388–418; Ciplet 2017; Warlenius 2018) based on the principles of historical responsibility and disproportionate use of a common but limited good (the ability of the atmosphere to accumulate carbon without causing dangerous climate change). A third reason developed countries ought to take on the most burdensome climate duties is based on a different principle — capacity to take on climate duties — given that their own industrial and economic development through long use of fossil fuels has also left them with a greater capability to both transition to clean energy and assist others in responding to climate change.

Another way to understand inequitable responsibility related to wealth has gained some prominence since the mid-2010s (Chancel and Piketty 2015; Oxfam 2015; Kartha et al. 2020; Chancel et al. 2021, chap. 6) by looking at emissions according to individuals' wealth. Between 1990 and 2015, just the richest tenth of humanity (around 630 million people) was responsible for a little over half (52 percent) of emissions linked to consumption, and the richest 1 percent alone (around 63 million people) for a whopping 15 percent. The poorest half of humanity, meanwhile, emitted just 7 percent of the total. On a per capita basis, the top 1 percent emitted more than one hundred times the poorest 50 percent. This twenty-five-year period, when annual emissions rose by

60 percent, also saw the same amount of greenhouse gas emitted as during the entire industrial period prior to it (Oxfam 2020b).

Accepting Fair Duties

In addition to determining fair duties, there is the matter of getting parties to not only agree to but also abide by them. The global negotiation process has never succeeded in setting the world onto emissions reduction pathways that are both equitable and capable of avoiding dangerous climate change. Nor has it been able to mobilize the funding that developed countries are supposed to be directing to developing countries to adopt renewable energy and adapt to climate change. Much of what has been directed is in the form of loans (rather than grants) and falls short, in particular, in terms of adaptation funding (Independent Expert Group on Climate Finance 2020; Oxfam 2020a).

In framing as a matter of climate justice the failure of rich countries to set and abide by ambitious targets, for both emissions and climate financing, our attention is drawn to possible reasons for this inability or unwillingness, which might include the absence of a strong and persistent enough democratic climate movement, dysfunctional democratic institutions, fossil fuel industry influence on democracy through contributions to proindustry politicians or climate deniers, resistance on the part of privileged populations or classes to giving up fossil fuels and attendant luxuries and hyperconsumption, or (neoliberal) capitalism and the underfunding of the public sphere.

This matter also demands consideration of the nature of each new global climate agreement, negotiated under the United Nations Framework Convention on Climate Change. The 2015 Paris Agreement was controversial because it did not legally bind countries to meeting their proposed emissions reductions or funding targets. After all, duties of justice are obligatory, not voluntary, and so require an institutional arrangement capable of enforcing them.

Who Will Be Impacted and Why?

Differential Vulnerability

A great range of factors influence communities' and individuals' vulnerability to the effects of climate change. At a macroscale, this involves countries' capacity to adapt to those effects, which is affected by economic development (itself affected by history) and geographical location (elevation from sea level, length of coastlines, etc.) At a mesoscale, these factors include the nature of the local economy. For instance, farming and Indigenous communities that are adapted to and heavily dependent on their respective local and traditional land bases can face difficult challenges due to climate impacts. At an individual level, intersectionality comes into play (Thomas et al. 2019), as gender, race, Indigeneity, class, degree of able-bodiedness, and more affect who is exposed and vulnerable to the effects of climate change and who has resiliency to recover from them afterwards. This injustice draws our attention to the people who are at the front lines of climate change and how structural inequalities exacerbate its impacts and dangers.

Colonial Legacies, Environmental Injustice

Another dimension of climate justice involves an inquiry into the historical factors that have led to exposure to the consequences of fossil fuel extraction processes and created vulnerability to the effects of climate change. Many of the communities on the front lines of extractive and refining processes and a crueller climate are there because of past injustices that remain unredressed.

As Roberts and Parks (2007, 104) put it, "The way a country is 'inserted' into the world economy bears heavily upon its ability to cope with climate-related disasters." They argue that the complex social, economic, and political structural changes imposed by colonial regimes to transform colonies into extractive economies based on a narrow range of low-value raw and barely processed materials left lasting legacies that increase the likelihood of being negatively impacted by climate change today. The legacies include high levels of rural vulnerability to disasters; exposure to flooding and storms in coastal areas; high economic inequality, which leaves the poor more exposed to the effects of disasters and less able to adapt and recover after they hit; and weak private property right regimes that prevent access to capital, insurance, and credit and that drive people into squatter settlements in areas at high risk of environmental disaster.

Journalist Christian Parenti (2011, 8, 11, 225–26) contextualizes climate change within the history of the Cold War and neoliberal capitalism. In the South, the North's Cold War proxy battles left in their wake armed groups, cheap weaponry, smuggling networks, and state corruption, while neoliberalism led to economic crises and inequality in developing countries and withered away governments' ability to pursue development goals. Climate change is now acting as an accelerant for the problems caused by this fatal combination, creating what Parenti calls the *catastrophic convergence.*

Potawatomi climate justice scholar Kyle Powys Whyte (2017) notes that settler colonialism has made Indigenous communities more vulnerable and less capable of adapting to the impacts of climate change in a number of ways, such as containing Indigenous communities to shrunken or marginal lands and politically disempowering them through the imposition of foreign governance structures. But too often unacknowledged is how the same settler strategies that have exacerbated climate vulnerabilities were also used to open traditional Indigenous territories up for extractive industries (including fossil fuels), deforestation, and agriculture — all major drivers of the climate change now impacting Indigenous communities. One powerful illustration of the role settler colonial history plays in the climate crisis and environmental injustice was provided by Aaron Carapella (2016), who has painstakingly put together a series of maps of the Americas showing the traditional territories of Indigenous Peoples. One of Carapella's maps shows just how extensively these territories have had to be crisscrossed by pipelines in order for the fossil fuel economy to exist. This colonial history explains why Indigenous communities have been on the front lines taking on the burden of leading so many resistance struggles against the industry's expansion (Indigenous Environmental Network and Oil Change International 2021).

Whyte (2019) further argues that climate justice cannot be properly advanced if it does not involve an understanding of the historical and contemporary injustices Indigenous communities experience due to the system of power and oppression created through capitalism, industrialization, colonialism, and militarism. State or corporate projects for lower-carbon or renewable energy can impact Indigenous communities by driving land theft or forcible displacement. And even a postcarbon world, if its institutions have not been decolonized, could retain and perpetuate colonial practices of forcibly taking control of resources and land for development.

Fossil Fuel Workers

In climate justice discussions, a careful and important distinction is made between fossil fuel *workers* and the fossil fuel *industry*. It is the latter that is targeted for its historical and political role in driving the crisis and preventing solutions from being adopted both in the past and today. Workers, meanwhile, are people who entered the fossil fuel industry in order to support themselves and their families in living a decent life. And they will be impacted by climate change because of the high-carbon-emitting nature of the industry they work in, which will have to be phased out in order to prevent the world from warming by ever more dangerous amounts. This is where discussions of a "just transition" for fossil fuel workers is raised (we return to this in Chapter 7).

What Is the Moral Significance of Climate Impacts?

We have a responsibility to investigate why climate impacts on people and ecosystems are morally significant. Through that investigation, we get a better sense of their moral weight, the urgency and priority that ought to be given to them. It is common to raise the ongoing and potential impacts of climate change in a way that lets their moral significance speak for themselves, to list the potential for food insecurity, water scarcity, displacement, destruction of property, loss of lives, etc. But we can go a step further. Highlighting the moral significance of these impacts deepens the perspective of the climate justice lens. It can elicit strong emotions of empathy or outrage. It exposes nuances that just responses must be sure to adopt. And it can justify escalations of measures taken to prevent further climate change.

Violence and Human Rights Violations

A first way to understand the deeper moral significance of climate change is to put it in terms of human rights. Human rights are meant to secure for every person a life of dignity and decency, a quality of existence that no one should be allowed to live below lest they lead a less than fully human life. To the degree that climate change threatens human rights, it becomes a matter of climate justice. Sea-level rise, temperature increases, extreme weather events, and changes in precipitation put all three dimensions of human rights at risk: civil and political rights (i.e., rights to life and property); economic, social, and cultural rights (i.e., rights to an adequate standard of living, adequate housing, food, water, subsistence, health, education, and cultural rights like those to world heritage); and collective rights (i.e., rights to self-determination and to the environment) (Schapper 2018; see also Alston 2019).

A related way of seeing the morality of climate impacts is to see them as a form of violence. To refer to climate violence evokes something more egregious, odious, and visceral than a violation of human rights. The writer Rebecca Solnit (2014) tells us "climate change is itself violence. Extreme, horrific, longterm, widespread violence." That violence is of an industrial and systemic nature, committed, intentionally and aggressively, against the poor in the case of fossil fuel companies. For Rob Nixon (2011, 2), climate change is a form of slow violence, "a violence that occurs gradually and out of sight, a violence of delayed destruction that is dispersed across time and space, an attritional violence that is typically not viewed as violence at all."

Perhaps another, starker way to put it is this: climate change is showing us who does and does not really matter. In 2013, just after the devastating supertyphoon Haiyan had struck the Philippines, the executive director of the Philippines Climate Change Commission said to fellow delegates at that year's United Nations climate conference, "Every time we attend this conference, I'm beginning to feel that we are negotiating on who is to live and who is to die." The power centres benefiting from the processes driving the crisis are able to do so because there is a class of human beings whose early and increasingly destructive experience of the effects of climate change do not matter much; to these power centres and those supporting or aligned with them, this class of humans are composed of *unpeople*. They perform an important function in maintaining the unjust status quo by being the kind of humans whose rights can effectively be made rescindable (Saad 2019a). Any climate justice response must be animated by a rejection and repudiation of this kind of human sacrifice.

Human rights violations and violence are similar in that they focus urgent attention on who experiences severe harm and how it can be stopped. And both prioritize a rapid, ambitious, and comprehensive response to climate change over other considerations. As Solnit (2014) states, "Once we call [climate change] by name, we can start having a real conversation about our priorities and values." This real conversation includes the question of whether and which drastic measures may be permitted in light of the severity of the crisis. As will be discussed in Chapter 10, this has long involved direct nonviolent action, increasing in recent years in the degree of disruption it seeks to achieve. And there is also now an emerging argument that the climate movement would be justified in engaging in sabotage and destruction of fossil fuel infrastructure (Malm 2021a), one that stems from the recognition of the human rights violations — and violence — that continued fossil fuel development will bring.

Though human rights generally define a minimum floor for well-being, we can go beyond and think about what makes life not only livable but enjoyable and purposeful. So in addition to thinking of the ways that climate change is eroding the pillars holding up human rights, we can consider the impact on those higher values that are endangered by climate change, whether conceived of in terms of capabilities (or human flourishing), freedom, sufficiency, economic growth and progress, or even pursuit of scientific knowledge and advancement.

Challenging Development Progress

A second way to understand the moral significance of climate change is to appreciate how it challenges international development progress. To be sure, development is a contested idea. But if it is taken to mean a process of improvement in the condition of people enduring deprivation so that they can pursue the human good (whatever that may be), then its prospects are seriously and massively threatened by climate change. The wide-ranging effects of climate change threaten to not only make development more difficult to achieve but to undo much of the progress achieved in alleviating poverty, reducing malnutrition, providing water and sanitation, and so on — all necessary to pursue the good. Consider how the United Nations Development Programme (2007, 1) put it:

> All development is ultimately about expanding human potential and enlarging human freedom. It is about people developing the capabilities that empower them to make choices and to lead lives that they value. Climate change threatens to erode human freedoms and limit choice. It calls into question the Enlightenment principle that human progress will make the future look better than the past.

Research suggests that climate change has already worsened global economic inequality, making developing countries poorer than they would have been without its occurrence (Diffenbaugh and Burke 2019). Climate change has the capacity to push 30 million to nearly 132 million additional people into poverty by 2030 by affecting agricultural productivity and food prices, exposing people to disasters, reducing labour productivity, and increasing morbidity, according to a World Bank estimate (Jafino et al., 2020).

Another crucial matter concerning climate change and development is how the "right to development" (Baer et al. 2008; Moellendorf 2014, chap. 5) can be protected and advanced while the world phases out fossil fuels. How are poor countries to provide energy for their economic development if not through fossil fuels? Will rich countries transfer to them the means to increase the availability of renewable energy or will they leave it to developing countries to figure out? This draws our attention to the importance of provisions in the current global climate regime that specify the nature and amount of North–South financial transfers, and whether the duties described therein are being met.

Burdening Communities and Threatening Ways of Life

A third way of seeing the moral impacts of climate change is through recognition that climate impacts create burdens for the vulnerable communities on its front lines. It is not that these communities are doomed or on the verge of extinction. (The Pacific Climate Warriors, a network of climate activists from several impacted Pacific islands, have taken a series of direct actions, including using traditional canoe flotillas to blockade coal ships, under the banner "We are not drowning, we are fighting!") These communities have agency and what is called in the literature *adaptive capacity* to moderate the effects of climate impacts, and their efforts might very well be successful, particularly if adaptation programs are participatory and well supported. However, the burden of adapting their societies to the effects of climate change, which these communities did not cause, is being imposed upon them by parties who are disproportionately responsible. And that burden grows greater as parties continue to fail to take climate action.

Speaking of Indigenous Peoples' resilience, Tom Goldtooth (2011a), executive director of the Indigenous Environmental Network, once said, "We have certain knowledge that we're able to adapt, but we should not be put into a position of forced adaptation or forced change.... Our forecast as Indigenous people is that, yes, we will survive, but we shouldn't have to go through all these difficulties.... We should not be put in that position."

Indeed, adaptation efforts might entail shifts away from traditional ways of life that are highly valued as a source of identity, pride, and cultural continuity or propagation. Where communities choose to migrate away from long-inhabited traditional territories, adaptation could mean that culturally cherished sites, important plant or animal life, or sacred burial grounds become difficult or impossible to access.

Whose Concerns Matter?

The effects of climate change are widespread and severe. They extend spatially and temporally, stretching far beyond the site and time from which the greenhouse gases that caused those effects were emitted. The decisions made now about how aggressively the world responds to climate change will have consequences for people's lives across the planet and far into the future.

Marginalization in Participation, Representation, Influence

One principle for evaluating the justness of the global climate agreement negotiation system is the degree to which it is participatory. Participation can mean ensuring that each national party has an equal voice, but it can go beyond this. In global climate negotiations, there is, historically, a pattern of the richest states holding overwhelming influence on the shape of the agreements produced. There ought to be a close relationship between the people who are most affected by climate change and their influence on how ambitiously and through what measures the world acts on the crisis. Furthermore, there will be parties not fully represented by their country-level negotiators — for instance, voices from grassroots organizations, peasant movements, vulnerable communities in the Global South, underserved communities of colour in the Global North, and Indigenous communities.

Democratic Deficits

The term *democratic deficit* is often used when ostensibly democratic institutions fail to generate political outcomes that would be expected under more participatory decision-making processes. Under liberal democracy, representatives make decisions in at least four key areas independently of the populace with regards to climate change that can prevent ambitious responses to the crisis: (a) whether to actually pursue policies on the crisis, (b) which policies to pursue, (c) the targets those policies are to meet, and (d) how they will be implemented, monitored, and evaluated and by whom. The population is generally not invited to submit its own solutions through any sort of participatory decision-making processes or given an opportunity to choose directly from a selection of solutions through a mechanism such as a referendum. It also has few means to guarantee that representatives, once elected, do in fact pursue the climate policies they have put forward in their political platforms. Rather, the population's role is confined to voting in periodic elections for parties whose platforms may or may not contain effective policies on climate change.

Political representatives' decisions on whether to act on climate change affects not just their own electorate but also populations and entire ecosystems across the world that experience its effects. These effects also extend temporally to future generations. The interests of this vast population have long remained excluded from the decision-making process.

Future Generations

Because a large portion of emitted carbon dioxide remains in the atmosphere for centuries, the effects of climate change will continue into the distant future. This means that questions of intergenerational justice are involved. The decisions made today have implications for what kind of world future people will live in. People born into a world with a destabilized climate face a host of potential hardships and suffering, which include being deprived of important human rights. Climate justice thus draws our attention to what duties we have now to preserve a safe climate as much as possible and to take the interests of future people into consideration (Page 2006). Much of the climate politics of the preceding generation treated the lives and well-being of members of future generations as effectively sacrificable.

The Nonhuman

The effects of climate change also impact nonhuman life. The 2019–20 wildfires in Australia were a horrifying reminder. A wwf-Australia (2020) report estimated that three billion animals were affected, adding, "It's hard to think of another event anywhere in the world in living memory that has killed or displaced that many animals. This ranks as one of the worst wildlife disasters in modern history."

Our dominant political systems have not been designed to take the interests of nonhuman life into consideration when making decisions. But if that other life matters, if we feel it is morally wrong to unnecessarily drive species to extinction or reduce complex ecosystems to ash, then we have additional interests that need to be represented in our political decisions.

What Is Driving the Crisis and Preventing Responses?

The Fossil Fuel Industry

It is impossible to talk meaningfully about climate justice without talking about the industry that is driving the climate crisis. To invoke climate justice with reference to the fossil fuel industry is to perceive the need to direct moral action against an entire sector of the economy. There are several reasons for this. First, and most prominently, the logic of its business model absolutely requires the destruction of the climate: Fossil fuel companies' value is based on the reserves that they hold and can sell in the market. However, they currently hold more in their reserves than can ever be used if humanity seeks to maintain a decent chance of survival, and their profit model demands that all of it is burned and that new stock is found to replace it (McKibben 2012, 2019c).

Second, and relatedly, fossil fuel companies have been aggressively pushing development of fossil fuel deposits that are more difficult to access and whose production carries greater risk of harms, and of infrastructure that exposes communities to risk of pipeline spills or exploding "bomb trains" as oil is carried by rail. All of this has involved putting pressure on governments to reduce environmental regulation to permit oil exploration, production, and transportation in high-risk areas.

Third, the industry has poured significant funding into obstructing climate policy by sponsoring climate change denier organizations (discussed in Chapter 5), lobbying decision-makers to adopt industry-friendly policies (Brulle 2018), and supporting profossil fuel politicians. Finally, governments continue to give subsidies to the industry. There are several major estimates for the size of contemporary fossil fuel subsidies, ranging from billions of dollars per year up to trillions if negative externalities are included (Bárány and Grigonytė 2015; Coady et al. 2015). These public subsidies go towards the industry driving the crisis and benefiting from environmental injustice at the same time that funds are desperately needed for investing in postcarbon energy systems and social infrastructure and for paying back a climate or ecological debt.

False Solutions

Critics charge that there exists a category of prominent responses that show greater concern for profit-making and economic growth than for climate preservation; they are attempts, in other words, to make climate solutions fit with the existing economic system rather than fit the economic system to the problem. The response that most commonly gets categorized as one of these so-called false solutions involves efforts to commodify emissions reductions through carbon offsetting arrangements, in which greenhouse gas polluters pay to have an amount of emissions equivalent to their own prevented or pulled from the air elsewhere in the world. Critics (e.g., Gilbertson and Reyes 2009; Gilbertson 2017) argue that there is a high degree of uncertainty as to whether offset programs actually eliminate emissions, and that these schemes incentivize land-grabbing and privatization of common resources, leading to dispossession and displacement of communities. Similar concerns have been expressed about Reducing Emissions from Deforestation and Forest Degradation (REDD+) (Goldtooth 2011b; Bayrak and Marafa 2016) and some biofuels (Rakia 2015). Geoengineering, the deliberate, direct, and often large-scale human intervention in the earth's climate system (and topic of Chapter 6), composes another set of responses that critics classify as false solutions.

Capitalism's Hegemony

Climate justice also interrogates the role of capitalism itself in perpetuating the crisis, not only by sustaining a reckless industry and promoting false solutions, but also by limiting the kinds of climate responses being seriously considered by governments and economic decision-makers. Neoliberal capitalism (the topic of Chapter 4) in particular is criticized for sharply reducing taxation on corporations and the economic elite and eschewing state economic planning and regulation, thereby weakening the public sector at a time when massive investments in renewable energy, public transit, and adaptation projects are crucial, as is the rapid phaseout of the fossil fuel industry (Klein 2014a). By interrogating how capitalism, neoliberal or otherwise, obstructs possibilities for fairer and fuller climate responses, climate justice draws attention to how the source of so many moral issues raised by climate change may be our reigning economic system.

A Rule of Climate Justice

As readers have no doubt observed, the issues revealed in exploring those five questions are many, complex, and severe, involving matters as crucial as whose lives and rights truly matter, who is most responsible for the emergency (and should therefore take on the greatest burdens), how legacies of unjust histories remain today in creating climate vulnerabilities, whether efforts to preserve hegemonic systems should be allowed to weigh down the response, and more. In raising these matters, the first facet of climate justice performs a crucial function: it makes it undeniable and unignorable that the climate crisis is a moral crisis, which cannot be fully understood or addressed without recognizing the various issues of justice raised by it. And note how these issues are not the kind that can be addressed with small, lifestyle changes or personal ethical consumption behaviours. Even today, individualist solutions get pushed, urging each of us to choose to take public transit (which is frequently underfunded and inefficient) or turn off our electronics when we are not using them. These are not solutions to an urgent and worsening moral crisis. Real solutions call for a *political* program of response that has morality at its core and as its overriding priority. Climate justice helps establish a rule: the less a political climate response acknowledges and is capable of addressing these issues, the less morally defensible it is.

A SECOND FACET OF CLIMATE JUSTICE: PRIORITIZING A MORAL RESPONSE

If the concern of a first facet of climate justice is to identify the moral issues raised by climate change and establish climate change as a moral crisis, the concern of a second facet is to shape and find ways of making real a response that prioritizes addressing those moral issues. As mentioned, this book cannot consider solutions to all of those matters. What it can do is lay out some general points to bear in mind while investigating the various ideological frameworks for understanding and responding to climate change.

A first general point emphasizes the prioritization of morality. If people and the planet matter, the climate crisis requires a response that far exceeds in urgency, scope, and ambition any solutions that are compatible with politics lacking a moral core. Responding to the climate crisis in ways that give primacy to competing concerns, ones that override and subordinate ambitious climate policies so that emissions reduction becomes just one concern among others, results in disastrously unambitious climate policies and leads in turn to climate injustices. They fail to put ethics first; moral outcomes may occur, but only potentially, a hoped-for side effect, not expressly prioritized or pursued.

The most urgent moral priority for a climate response is to offer a means of achieving highly ambitious emission reduction targets. The less concern a climate response shows for such targets — targets that preserve as much as possible of the climate our societies are adapted to — the more strongly we should condemn that response because it is implicitly willing to burden, risk, or even sacrifice individual lives, communities, and ecosystems in order to prioritize some other logic.

The problem is that there are any number of such competing considerations, including the preservation of economic growth, adherence to the neoliberal orthodoxy of minimal taxation on wealth and corporations and minimal public spending, protection of a domestic fossil fuel industry, and avoidance of political controversy engendered by disrupting the status quo. Climate justice seeks to prioritize moral concerns above others where they are in contestation.

Certainly, prioritizing moral concerns is not always a straightforward task. Proponents of those competing rationales often argue for them in terms of the greater good they promote. Sometimes, the defences given are simply dishonest attempts to confuse the debate on climate change. This occurs, for instance, when climate change deniers assert that transitioning away from fossil fuels would necessarily deprive communities in the Global South of energy necessary for their development — a false dichotomy since international support in building clean energy infrastructure would allow developing countries to leapfrog over fossil fuels. But in other instances, the problem is more fraught. Back in 2014, under fire for ignoring the climate crisis and seeking to turn the country into an energy superpower by developing the vast oil reserves in the Alberta tar sands, then Canadian prime minister Stephen Harper said the following:

> It's not that we don't seek to deal with climate change, but we seek to deal with it in a way that will protect and enhance our ability to create jobs and growth, not destroy jobs and growth in our countries. And frankly, every single country in the world, this is their position. No country is going to undertake actions on climate change — no matter what they say, no country is going to take actions that are going to deliberately destroy jobs and growth in their country. We are just a little more frank about that, but that is the approach that every country's seeking. (CBC News 2014b)

According to Harper, economic growth should be a greater priority than climate action, Canadian jobs more important than the lives burdened and lost across the world as a result of unleashing the vast carbon stores buried dormant in the Alberta tar sands, as he was attempting to do. There is every reason to contest claims and framings of this sort. Indeed, as Chapter 8 shows, there are some who argue that perpetual economic growth is not only environmentally destructive and impossible on a finite planet but also unnecessary for human well-being. It would follow, then, that we need not prioritize it, which could allow us to more aggressively forgo the large amount of fossil fuel energy devoted to economic activity conducted only in the name of more growth. But there are also those who argue, in turn, that economic growth is a necessary condition for human well-being and progress, and that to hamper it in the name of climate action would give rise to unnecessary economic hardship or worse.

Stirring up this kind of debate about what we prioritize is the point. And it's this debate that leads to a second general point. Climate justice demands that we do not take the contemporary political and economic order for granted. That order is not self-justifying, particularly if preserving it stands in the way of averting the worst effects of the climate crisis. If we find that there are parts of this order that are preventing us from taking action at the scale and speed morality demands, then those parts should be open to critique and change. And because a climate response requires changes to the way society functions, there arises an opportunity to ensure that those changes are made in ways that not only preserve the best parts of society but make a more decent one that better promotes the human good. System change to fight climate change in a moral way has to be seriously considered.

For instance, at a historical moment when capitalism's neoliberal form has entailed economic crises, sharp economic inequality, austerity in public spending, and increasing precarity in working life, its failure to address climate change makes it, in the eyes of its critics, a system twice damned. Not only might alternative economic arrangements better address climate change, but they can also create fairer and more meaningful ways of life. An insistence on resolving climate change through neoliberal capitalism would thus be irrational and immoral. Similarly, following long histories of colonialism and imperialism filled with atrocities and injustices visited upon Indigenous communities in the name of capitalism, resource extraction, modernization, and growth, is it not essential to ask how we might live under a different system that eliminates these colonial drives? At the same time, drawing attention to the system-changing potential of climate justice raises questions over how much of the existing order can be challenged in time to address the crisis.

Third, a climate justice lens should also draw our attention to the ways in which responses have been conceived and developed, by whom, and in what interests. We can conceive of responses as falling along a spectrum based on their origin, with *technocratic* on one end and *grassroots* on the other. Technocratic responses take for granted that any climate policy should avoid altering the existing system as much as possible. Or put another way, they leave existing political and economic arrangements largely intact as a condition for taking climate action; climate action that is too disruptive to that status quo is rejected outright. They originate from policy experts that have been commissioned by governments or are writing with reigning policymakers in mind. To the extent that there are moral concerns contained within them, they compete against but do not override other concerns, particularly for nondisruptive change, which can involve all sorts of concessions, to the fossil fuel industry driving the crisis, for instance. They are attempts to innovate within (and protect) orthodoxy and the status quo. Grassroots responses, on the other hand, emerge from a variety of communities engaging in political struggles. They have no direct legislative power or strong influence, but do make the strongest moral demands; indeed, it is moral concerns that drive grassroots action, which arises from a sense that those in charge are failing to do what is right and necessary.

The question of representation of the underrepresented and voiceless raised above is important here. There is a commonsensical way of seeing democracy as merely a process in which people choose from a set of predetermined choices. Elections of government parties that have presented policy platforms form probably the most significant experience of this. But this is a very shallow form of democracy. Much deeper and fully democratic solutions find ways to consider or represent the concerns of those not involved in but nevertheless impacted by a decision, whether elsewhere in the world or forward in time. As we consider the ideological frameworks for responding to climate change, readers should question how well the respective responses would address the concerns of those communities on the front lines of worsening climate impacts and of future generations.

GROUNDWORK LAID

We have now covered the essential groundwork elements for this book: political ideology and climate justice. The following few chapters highlight how political ideology creates competing frameworks for responding to climate change, specifically in relation to reducing greenhouse gas emissions. The chapters open on the core beliefs these ideologies hold and why, then show the climate response consistent with them. Towards the end of these chapters, we examine the frameworks through a climate justice lens and highlight potential moral issues that arise from attempts to address climate change through them. My hope is that, using the discussion about the two facets of climate justice, readers will expand on those critiques wherever I have failed or been unable to include something.

PART 2

....

THE SYSTEM-PRESERVING FRAMEWORKS

4

NEOLIBERALISM AND THE WORLD-SAVING MARKET

THE ENGINE MAKES A *healthy, confident electric whirring as the car accelerates away from the rental tenements and towards the gleaming city centre. The young woman sits back as the auto drive takes over, communicating with all the other self-driving vehicles throughout the route to coordinate a rush-hour traffic flow smoother than human minds alone could achieve. "News," she says, and the windshield darkens momentarily, blocking the outside world, before displaying a menu grid of moving images with titles like "New study: Major wearable tech brands using African sweatshop labour," "Wealth inequality reaches new historic highs," and "Space-mining magnate is world's newest trillionaire." She looks at the one titled "More of Miami now permanently underwater" and blinks twice to select it, watching as its borders expand to fill the windshield screen. The voiceover talks of how emissions released generations before warmed the earth, committing the planet to sea-level rise for centuries to come.*

Why did the people of the past emit these planet-warming gases so recklessly? She remembers her grandparents, shrugging sheepishly, explaining they didn't really have a choice as they told her of a time when vehicles were filled with a liquid fuel to power their propulsion.

"Where did the fuel come from?" she asked.

"Gas stations," they told her, but she didn't really understand. And neither did she really understand why people once fought wars over the stuff or how political parties with close ties to the "fossil fuel" industry could lie about what using it was doing to the world. It all seemed so primitive to her. And in a way, this was literally true: that stuff, she would one day learn, was the leftover matter of eons-dead organisms, pressed together, inexorably, and condensed in the depths of geology and time. Set alight,

the dense and ancient chemical bonds severed explosively, the combustions harnessed to drive motors, the waste product set free in the air to warm a planet.

Perhaps to prevent her from placing too much blame on them, her grandparents insisted they really had been ready to make sacrifices to protect as much of the future world as they could. In their youth, they supported politicians willing to make hard decisions balancing the economy and the environment. Prices went up for a bit on fossil fuels and all that relied on them — which was basically everything — until finally everyone made the switch away from them. No, it wasn't fast enough, but, as far as they remembered, there was no alternative.

The vehicle stops at her destination, the first of two contract jobs she is working today. A kind, if sterile, electronic voice tells her how much the ride-sharing app has automatically charged to her credit account before the car speeds away through the city to take the next rider wherever they need to go. She would love to spend some time learning more about the 2020s, the era of her grandparents, when the very world was at stake. But until she found something full-time so she could pay off her student debt and maybe one day afford a house (who was she kidding?), there were precious few hours for anything like that.

NEOLIBERAL IDEOLOGY AND SOCIETY

The gleaming neoliberal future described above is, potentially, one world in store for us. It's a world of high tech and low equality, overwhelming work and underwhelming prospects. And it would not take much to usher it in aside from leaning into the trajectories of today's dominant political and economic order: neoliberalism.

In order to understand the ideology of neoliberalism, we need clarity on what liberalism is. Liberalism is a notoriously contested concept with no shortage of meanings; indeed, over time, some quite different political formations have been referred to as liberal. There is always, therefore, some controversy inherent in defining the term. In this book, liberalism refers to a political worldview that believes the economic and political institutions of capitalism and representative democracy are the ones best capable of creating an ideal society, one that is characterized by the maximization of individual freedom and equality, the promotion of rationalism and prosperity, and continuous progress in the human condition. (To be sure, this sense of liberalism is a contemporary one, skipping over the tensions and contradictions historically involved in protecting the liberties of the individual — as well as the privileges of the minorities in the upper classes — on the one hand, and promoting the power of the masses to democratically shape society even in ways that reduce privileges of individuals in the ruling classes, on the other hand [Meiksins Wood 1995, chap. 7].)

While adhering to this general worldview, liberals may disagree about the particulars. The key values of freedom, equality, and progress can be interpreted very differently. And any liberal society has to decide how the state will administer capitalism. Should the state harness and moderate capitalism to promote a society characterized by relative economic equality and human flourishing? Or should the state promote the interests of the capitalist class by, for instance, imposing the imperatives (or "logics") of the market — competition, deregulation, privatization, enclosure, and commodification — over ever more parts of society?

We can think of these matters of disagreement as "shearing forces" within liberalism simultaneously pushing it in different directions. The result is a range of liberalisms. This book focuses on two important variants. The most left-leaning of the two, social democracy, is explored in Chapter 7. For now, suffice to say that social democrats believe capitalism promotes prosperity and progress, but worry that freely unleashing capitalist market logic and creating a context of intense intrasocial competition to accumulate economic power and maximize profit can have enormously burdensome, and even destructive, effects on the larger society. Neoliberalism, the topic of this chapter, is the most right-leaning and dominant of the two variants. For neoliberals, the very market logics that social democrats want to restrain are those we ought to set free.

The ideological rationale typically given is that the market functions best when it is free of distortions, and that a well-functioning market will maximize economic growth and the social good by giving all actors the information needed to make rational economic decisions. But for this to occur across a society, the market must be left as much as possible to its own devices. The more a government attempts to manage the capitalist market and its effects, the more it distorts the workings of the market, and the more inefficiently a society's resources are used.

In order to allow the market to operate more freely, societies shaped according to neoliberal ideology take on some prominent characteristics — ones that, critics urge, have made it difficult to address climate change over the last generation that neoliberalism has been hegemonic. The first is a preference for economic deregulation. Proponents of state regulations attempt to promote and protect the social good by placing conditions on (and therefore restricting) how freely actors may operate in the marketplace in order to gain competitive advantage and maximize profit. For example, minimum-wage regulations specify the least an employer can legally pay workers, rent controls specify how much rent can be raised in a given period, environmental regulations specify how pollution must be dealt with (if at all), and so on. But neoliberals argue that reliance on regulations, even in the hands of well-meaning governments, risks burdening economic actors, making them uncompetitive or unprofitable in a global marketplace, or passing costs onto consumers. Instead, industries are given more freedom to self-regulate — to ensure, under their own guidance, that their practices will not be harmful to the social good.

A second characteristic of neoliberal societies is minimal corporate and wealth taxation leading to public austerity. Giving the market freer rein involves reducing taxation, particularly on corporations (in order, neoliberals argue, to improve competitiveness) and on the wealthy (in order to fully liberate their entrepreneurialism and encourage investment). The natural result, as budget cuts proliferate, is a public sphere with fewer resources for governments to invest in formerly free or affordable services. Neoliberals argue that the resulting public austerity is part of creating a more economically efficient and rational society. Rather than being lavishly resourced with taxpayer wealth, the public services that governments play a role in providing are forced to be smarter, to do more with less by shifting their operation models to approximate for-profit businesses. The university is one prominent example, where, as public funds dry up, teaching duties are increasingly conducted in large classes by contract faculty and graduate student teaching assistants while tenured positions become increasingly rare, tuition costs rise, and lucrative markets for international students are pursued.

Closely related is a third characteristic of neoliberal societies: increased privatization and commodification. This is because another economically rational response to public austerity is to sell off and privatize government services — that is, to have them provided by for-profit actors as commodities. Sectors to privatize can include telecommunications (e.g., internet and cellular service providers, publicly owned media), utilities, education, social housing, pensions, health care, child care, elder care, parks and leisure, transit and transportation, and even security and incarceration. In addition to the commodification of public services, neoliberal societies commodify goods not normally sold for profit and give a role to government in creating well-functioning markets for them. It's a strategy that is often used for attempting to respond to some social problem without turning to government control. Pollution permits, or purchasable rights to emit destructive substances into the environment, are one example discussed below.

Since the 1980s, neoliberalism has been the dominant political ideology throughout the Global North and most of the world, shaping its societies for a generation. In that time, it has been subject to no shortage of sweeping critiques for a host of reasons: driving obscene wealth concentration, social immobility, and corporate domination (Chomsky et al. 2017); creating persistent conditions of work precarity (Standing 2011); being imposed opportunistically over people against their will in moments of shock and crisis (Klein 2007; Loewenstein 2015); increasing the frequency of financial crises while weakening the capacity of society to respond to their effects; undermining the essential elements of democracy by remaking the function of states and the behaviour of individuals so they act as little more than competitive, value-maximizing firms (W. Brown 2015); and being a transparently superficial theoretical justification for a reactionary political project intended to restore wealth and power to the capitalist class following the postwar era of redistributive Keynesian economics (Harvey 2005).

When it comes to climate change, critics (e.g., Klein 2014a; Fremstad and Paul 2022) have observed how neoliberalism's insistence on deregulation, public austerity, and privatization left it without the means to address the crisis boldly enough. And indeed, the crisis created a tension in maintaining the legitimacy of neoliberalism. A fanatical trust in free markets to solve the issue would be irrational; there is nothing about the functioning of markets themselves that would drive a rapid switch away from fossil fuels at anything like the pace required. But to depart from market logic too much would fly in the face of the accepted orthodoxy for the last generation.

NEOLIBERAL CLIMATE SOLUTIONS

For a long time, the main solutions the neoliberal order was willing to offer were campaigns to promote individual actions — "ride your bike," "shut off your lights," "take public transit" — that alone are insufficient to significantly reduce emissions. And so some finessing was in order. This section looks at the main policy responses that emerge from this framework including carbon pricing, regulations, public support, and research and development.

Carbon Pricing

"The greatest and widest-ranging market failure the world has seen."
If there was a watershed moment marking the point when the neoliberal order could find a way to come to terms with the climate crisis, it was in 2006–07 when these words began appearing in media headlines and think pieces. This is the way that Lord Nicholas Stern (2007) famously described the problem of climate change in his landmark report for the UK government on the economics of climate change.

In essence, Stern was saying, the market is lying to us about the true costs of fossil fuels. Whenever we pay for electricity generated by coal or natural gas, heat our homes with natural gas, or fill our tanks with gasoline or diesel, the price takes into account the costs of discovering, extracting, processing, transporting, and distributing those fossil fuels. But that price is also committing a lie of omission, leaving something out that, if we knew about it, would lead to less demand for fossil fuels. What it's not informing us about are the high costs to society of using these fuels.

In economic parlance, this is known as a *negative externality*, where the undesirable consequences of a market transaction (here, buying fossil fuels in order to burn them for energy) fall onto other parties not involved in, or external to, that transaction. Externalities are common and not restricted to the matter of climate change. When we pay for clothing that has been made in a country with low environmental and labour standards, the price does not include the impacts to communities situated close to the factories — impacts that might include pollution of rivers or local air. The clothing can be cheap for us to buy (and for major corporations to produce) because locating factories in these places allows producers to displace the costs of dealing with the consequences of production onto others.

In the case of fossil fuels, it is almost as though an invisible portion of the price we pay is being subsidized by those who have to foot the bill for the fuels' impacts, like subsistence farmers in the Global South using their scarce resources to fight off an encroaching desert or people living on low-lying island states paying to relocate away from rising seas. There is, in other words, a social cost of carbon. And as long as that cost fails to figure into the price paid for fossil fuels, an economy will demand more of them than it would if that cost were included.

Cap-and-Trade

There are two ways of putting a price on carbon in order to "internalize" the externality. The first, cap-and-trade (or emissions permit or emissions trading), is the more complex. In principle, it works like this: Governments issue permits to companies, allowing only a limited amount of carbon to be emitted in a given period (this is the "cap" in cap-and-trade). High-carbon sectors of the economy (e.g., power generators, oil refineries, airlines) must acquire permits sufficient to cover their emissions for that period or face some form of penalty. Should a firm reduce its emissions below the level allowed by its permits, it will hold an excess supply that it can sell to others (the "trade" part of cap-and-trade) who are in danger of incurring a penalty for exceeding the emissions covered by their permits. Cap-and-trade thereby produces assets — in this case, the right to pollute the atmosphere — that can be commodified. Achieving emissions reductions depends on the carbon cap being set below the business-as-usual level of emissions, creating a scarcity of permits in the market that then determines the carbon price. The rarer the permits, the higher the price of carbon.

As time goes on, the total number of permits that governments issue falls. Actors in high-carbon sectors are thus incentivized to innovate or find low-cost means of decreasing emissions in order to (a) avoid needing to purchase increasingly scarce (and therefore increasingly expensive) permits in the future and (b) be able to sell any excess permits they hold to firms who have failed to acquire enough to cover their own emissions.

Supporters note that this system allows greater flexibility in meeting climate goals because actors can seek out the cheapest and most efficient ways of lowering emissions. If properly implemented and enforced, cap-

and-trade also guarantees a specific reduction outcome because the reduction amount is set out in advance. Cap-and-trade furthermore avoids the imposition of new taxes (like carbon taxes, described below), which tend to be unpopular with voters. Finally, the sale of permits through public auctions opens the opportunity for governments to raise revenue for green initiatives, as the state of California and the province of Quebec do.

Detractors, meanwhile, highlight the added costs of establishing a bureaucracy to govern the cap-and-trade system. Additionally, the way that costs are imposed upstream on firms in select industries rather than on consumers at the point of sale can mean that the wider public does not process, psychologically, that their high-carbon purchases are contributing to climate change.

The last chapter raised another main target for criticism when it comes to cap-and-trade. Carbon offsetting occurs when carbon-emitting parties make a payment or investment that enables some operation to prevent an amount of emissions equivalent to their own from being added to what is in the atmosphere. The polluter party thereby earns an emissions credit similar to the emissions permits issued by governments. The argument for offsetting is that because greenhouse gases are well mixed in the atmosphere, it does not make a difference where they originate; therefore, entities and individuals need not eliminate their own emissions, and can instead compensate for them using the cheapest means to do so wherever the option is found in an emerging global offset market. In this way, offsets offer flexibility. And these offsets can be used in either compliance markets (where actors must comply with government legislation to achieve some emission reduction goal) or voluntary markets (where businesses, financial institutions, governments, or individuals are seeking to improve their sustainability independent of any adopted legislation).

But there are some issues that arise. Arguably the biggest is the problem of *additionality* — that is, the matter of whether a carbon offset project actually prevents emissions. The investment by a party seeking to compensate for its emissions is what is supposed to allow an offset activity to proceed, so if a given amount of emissions would have been avoided or withdrawn even without the incentive created to earn a credit, then there is no additionality. For example, in a 2016 study Cames et al. found that 85 percent of the 5,500-plus carbon offset projects they surveyed had a low likelihood of leading to additional reductions. Many were energy-related projects that would have probably gone forward anyway because they were profitable even without investment from rich countries. In other words, the rich-country parties probably did not actually offset their emissions but received credits as though they had (see also Farand et al. 2022). The danger with offsets is that suppliers are incentivized to sell credits but not necessarily guarantee additionality, while buyers are incentivized to purchase (cheap) credits without being certain of their additionality.

Carbon Taxes

Carbon taxes are the second and simpler instrument for applying a carbon price. A government decides on the degree to which carbon emissions should incur an economic cost (the cost is usually levied at dollars per tonne of carbon dioxide). It also decides which sources of carbon emissions incur this cost, whether the cost will rise, and on what schedule. Thus, governments have a way of moderating the influence of carbon pricing so as not to disrupt an economy's growth or competitiveness.

A major question is what to do with the revenue collected through a carbon tax. There are two general approaches: revenue-neutral and revenue-generating. The first approach can come in a couple of forms. One is to reduce taxes (e.g., income taxes) by an amount equivalent to what the carbon tax brings in. Another is sometimes called fee and dividend, where the amount of revenue taken in each year by a government's carbon tax (the fee) is returned back to individuals or households in equal amounts (the dividend). Proponents believe that this is the most politically palatable approach, arguing that even though everyone receives the same amount of money, it ends up being a progressive tax, one in which lower- and middle-income earners actually receive more through the dividend than they pay. This is because richer households and corporations, with their high-carbon practices, end up paying so much.

With revenue-generating carbon taxes, governments use the revenue for a specific purpose. That might include investing in renewable energy projects or in public transit initiatives. More radically (and not generally advocated for under this framework), the revenue may be put towards repaying a climate debt by transferring funding to developing countries for their own emission reduction and climate change adaptation efforts.

New Price, New Rationality

Adding a cost to greenhouse gas emissions will change the background conditions against which market actors make rational, self-interested decisions by raising the cost of fossil fuels relative to cleaner energy alternatives. Once the new rules of the game are established, rational market actors will adjust, the smartest investors will foresee the gains from the transition, and there will be vested private interests in carbon-free technologies. In short, carbon pricing creates a situation in which renewables become increasingly acceptable, and the switch to them increasingly rational, as long as they outcompete fossil fuels in the market. Amid the dramatic plummet in wind and solar energy costs, as conditions for a transition without disrupting capitalism become increasingly favourable, the neoliberal system seems readier than ever to deploy moderate carbon pricing, particularly as studies show it will have little impact on economic growth.

Carbon pricing has now become the choice climate policy in the world. The number of carbon-pricing initiatives that have been implemented or scheduled to be implemented has jumped rapidly since 2010, when there were just nineteen. By 2015, the number had doubled to thirty-eight and by April 2022 had more than tripled to sixty-eight altogether (World Bank 2022). In January 2019, a signed statement — billed as "The Largest Public Statement of Economists in History" — appeared in the *Wall Street Journal* calling for the United States to adopt a fee-and-dividend carbon-pricing arrangement (Climate Leadership Council 2019). Indeed, carbon pricing has become so mainstream that even major oil and gas corporations are (ostensibly) supporting it.

Carbon-pricing policy is intended to do the heavy lifting in realizing the emissions reductions schedules set by policymakers. These schedules can come in the form of emissions reduction goals relative to a baseline year, or, as is becoming more common, in the form of a deadline by which a country is carbon-neutral. These two levers of control — over emissions reductions targets and timelines and over prices on carbon — allow policymakers to shape climate policy cautiously to avoid affecting economic growth and to respect the pace of change that market actors can accommodate. The stage is thus set to hand over an extremely large part of the climate response to business ventures. Indeed, there is now a clean technology press, composed of groups like CleanTechnica and Bloomberg

New Energy Finance, which regularly points out the enormous business opportunities that renewable energy and green tech are opening in the fabric of twenty-first-century capitalism. In the wake of US president Donald Trump's announcement in 2017 that America would withdraw from the Paris climate agreement, domestic commentators (e.g., Romm 2017) quickly pointed out how unwise it was to hand over the keys of private clean energy growth, one of the major motors of contemporary capital accumulation, to competitors in Europe and China.

But so far, the actual price of carbon that governments have been willing to impose has remained below what economists say is required to achieve the kind of emissions reductions necessary to meet the Paris climate agreement goal of keeping temperature rise below 2C. According to the High-Level Commission on Carbon Prices (2017), emissions need to be priced at $50–$100 per tonne of carbon dioxide equivalent by 2030. But by April 2022, less than 4 percent of priced emissions were in that range. Furthermore, only around 23 percent of global emissions were covered under carbon-pricing initiatives (World Bank 2022).

Regulations

As noted earlier, neoliberalism generally eschews regulation, and yet regulations compose the other major set of climate policy instruments acceptable under the neoliberal framework. An important exception is being made in the case of climate change *in order to get the market to function properly*; proposed regulations remain far away from dictating a time by which the fossil fuel industry must produce its last barrel of oil. In an indirect way, regulations also put a price on carbon. The price that firms incur is through the adjustments required to adapt their products to be compliant with new climate regulations or else be unable to sell those products in the market.

One prominent example of the use of regulation in recent years comes from initiatives by governments to set target dates after which vehicles powered only by internal-combustion engines can no longer be sold. Automobile producers interested in remaining in the market are thus incentivized to make the investments needed to retool and reorganize in order to produce competitive electric (or, in the interim, hybrid gasoline–electric) vehicles.

This sort of regulatory approach finds support among some commentators for avoiding the political contentiousness arising from more overt carbon-pricing measures. For one, carbon pricing is highly susceptible to political mischaracterization and can unite powerful opposition. The average voter can be easily turned against policies to raise taxes. Australia initiated one of the earliest carbon-pricing policies in the world under Julia Gillard's Labour government, only to lose power to the Conservative party under Tony Abbott, who campaigned successfully on a platform to "axe the tax." In the late 2010s, a number of Canadian provinces saw right-wing governments come to power under promises to fight the federal Liberal government's carbon-pricing initiative. In Washington state, citizens have been convinced twice, once in 2016 and again in 2018, to reject ballot initiatives to institute a carbon tax, the second time with Big Oil working to pour significant funding into the "no" campaign (Roberts 2018a).

It is for precisely these reasons that some thinkers prefer regulations over carbon pricing. These regulations can exist along a spectrum (Jaccard 2020, 109–14). On one end are prescriptive regulations, which call for a specific technology to be used or a precise emission level for each individual actor or firm. On the other end are flexible regulations. One example is a renewable energy portfolio standard, where a particular jurisdiction has to have a certain amount of its electricity generated by renewables. Unlike with prescriptive regulations, not every actor has to have the same amount of renewable energy generation. Individual energy suppliers that do not meet the average specified in the regulation can buy credits from others that exceed the average, granting a subsidy of sorts to cleaner energy suppliers while raising their own costs. Similar logic could apply to zero-emission car manufacturing standards (manufacturers that do not produce the legislated percentage of zero-emission cars can buy credits from those who exceed the percentage) and low-carbon fuel standards.

Like carbon pricing, regulations are also challenged by that weighty neoliberal presence. But the market and its actors are nevertheless retained here as the protagonists in the case of flexible regulations. Nothing about the approach requires a wholesale departure from the sphere of neoliberal influence.

Research and Development

Briefly, there is one other important market failure to correct. Research and development of unproven technologies and innovations tend to experience underinvestment from the private sector due to low returns on investment. Most early trials and tests fail to produce anything profitable, while certain breakthroughs become public knowledge, meaning that, even if society benefits, private investors do not. To promote rapid advances in postcarbon technologies like renewable power generation, storage, and transmission and zero-emissions liquid fuels, coolants, and concrete, governments will need to step in to make investments that are underprovided in the free market.

THIS FAR, NO FURTHER

Through those policy tools, the neoliberal framework reduces the climate crisis — a complicated, multifaceted "wicked problem" — to a much simpler matter. At the root of climate change is a market failure, and thus the crisis is a (relatively) simple problem of mainstream economics, public policy, and — once the market failure is addressed — technology. It is a matter of incentivizing proper economy-wide behaviour. Just as there is a scientific consensus on climate change, there is a neoliberal consensus on climate policy: political representatives initiate carbon-pricing measures or regulations that will trigger technological innovation in the renewable energy sector, whose products can in turn outcompete the now increasingly more expensive carbon-intensive alternatives, leading to an efficient switch to a carbon-free, capitalist economy. Our lives will be marked by relatively little disruption in the transition to a postcarbon world. For most people, it will be one occurring largely in the background, as decisions towards lower-carbon lifestyles are subtly nudged by market signals. If events unfold according to plan, fossil fuels are swapped out for clean energy that can, in perpetuity, sustain capitalist production and growth.

Indeed, this pragmatic straightforwardness is one reason that, the argument goes, climate policy does not have to be tied to demands for justice. There is no need for any nonsense about overhauling the economic or democratic systems outright as more progressive and radical frameworks would insist. There is a hard-headedness about the need for compromise and gradualism. And so, another important feature of the neoliberal response to climate change is its role of gatekeeper, closing off pathways to more politically progressive and radical agendas that would leverage the crisis to engage in leftist tinkering with the carefully calibrated arrangements that neoliberals believe are required for human well-being.

Starting around 2015, we saw this attitude in reactions to the sudden rise of proposals for more far-reaching climate responses, whose proponents believe that the rules and power structure of the existing neoliberal order prevent us from achieving sufficiently rapid decarbonization. The urgency of the crisis makes it necessary, these progressives and leftists argued, to turn to much bolder alternatives, which tend to include (a) the foregrounding of justice-based demands (e.g., employment and housing guarantees, redress for historical racism, expansion of affordable education and health care services) that can mobilize a large progressive democratic movement and, more radically, (b) some form of engagement with or transcending of existing neoliberal capitalism.

Let's take a close look at why the neoliberal framework rejects these kinds of alternatives, beginning with its opposition to comprehensive justice-based solutions. To defenders of the neoliberal climate response, those solutions are the very opposite of what is needed: the urgency of the climate crisis is precisely why the climate response should be shaped within the system and kept as directly related to emissions reductions as possible. Such demands, under the neoliberal framework, are tangential at best, and at worst so politically contentious as to be detrimental to winning a realistic climate response (Jaccard 2020, 224–38), which simply requires the right combination of market-oriented policies (Chait 2015; Rand 2020). It is already challenging enough to get the average person to accept mechanisms like carbon pricing and regulations, especially with politicians further to the right holding out the alluring promise that we need not do much of anything at all about climate change. So why complicate it with the divisive politics of social justice? The key to committing society to any kind of climate action is to aim strictly for the politically doable.

Anything more ambitious would alienate the bulk of voters or, worse, result in backlash that would work in the favour of political forces hoping to rescind climate legislation altogether. Much of this stems from pessimism in right-leaning forms of liberalism about the possibility of

far-reaching, progressive change in a society characterized by a plural-
ism of political concerns, most of which are best attended to by a politics
of reasonable compromise. Picture the electorate as a semisolid gelati-
nous mass shaped like a bell; on the far right and far left the mass thins
out, representing the comparatively small number of people believing in
the policies at the ends of the ideological spectrum. But in the centre it
bulges upwards, representing the large number of people holding moderate or centrist beliefs. The problem is this: the mass shifts only so far in
response to new ideas that depart from the status quo. Leftists may make
all sorts of claims about the good that will come, theoretically, from radical change, but they are ignoring that the surest path to electoral victory
is to play to that sizable swell of moderate political opinion in the centre
largely resistant to being pulled far to the right or left of the bell.

Indeed, people dedicating their efforts to winning justice-based responses might even be engaging in a destructive form of self-indulgence
manifesting in a refusal to support politicians whose policies are im-
perfect but who are nevertheless sincere about tackling climate change
(Jaccard 2019); withholding electoral support only plays into the hands
of politicians further to the right with no interest in climate action. The
2016 Washington carbon tax vote is one example sometimes given to
illustrate the irrationality of the climate left. There, the climate justice
movement urged voters to reject a ballot initiative that could have seen
a carbon tax put into law — the objection was it would have been revenue-neutral, and therefore would not have raised financing for justice-based climate projects. Neoliberal voices (e.g., Pinker 2018, chap.
10) point out how opposing the legislation put the left on the same side
as the fossil fuel magnate Koch brothers. Other leftist positions are similarly criticized, like the rejection of nuclear energy, which provides a
source of electricity that is abundant, safe, stable, and largely carbon-free
(emissions are involved in mining for uranium and separating the ore
from rock).

As for transcending neoliberal capitalism, this is also ill-advised. For
one, defenders of the status quo argue, it would be impossible to do in
time (Pesca 2019). And, second, it is not desirable. Neoliberal critiques
of the responses to the climate crisis proposed from further to the political left (like the Green New Deal or, before it in Canada, the Leap
Manifesto) were occasionally made in terms of defending the productivity, wealth, and innovation of the existing capitalist system required for

human welfare (Homer-Dixon 2016). And so there is a vigilant wariness towards leftists attempting to change the climate crisis into something more than it is, to turn it into yet another attack on the capitalist system by people who do not appreciate that it is essential for promoting prosperity, innovation, and progress (Pinker 2018; Chait 2019).

So if it is not the capitalist system itself that is at the root of the problem, as more radical voices would insist, what else could be preventing climate action? Human psychology, individual and social, is one explanation that is compatible with this framework and precludes further inquiry into how the contemporary economic system might be to blame.

First, climate change is a strange kind of threat. People have evolved to focus on and react quickly to clear and present threats, both in the form of dangers to our individual selves, like a predator or attacker approaching us, or to our societies, like pandemics, terrorist attacks, or invading armies. These kinds of threats trigger us to rapidly take protective or defensive measures (or to accept them when imposed by authorities). But the nature of the climate crisis is such that it fails to activate the kinds of reactions those more immediate dangers do; it is not the kind of problem humans faced in our evolutionary past. Though people may speak of a climate *crisis*, we don't tend to feel a rush of blood or surge of adrenalin when thinking about it. The result has been delay and deprioritization.

Second, the processes driving the emissions of greenhouse gases involve a Faustian bargain in which longer-term consequences are brushed aside in order to reap substantial, convenient, and immediate rewards in the form of the immense energies that come from fossil fuels and make the benefits of our modern civilization. It is not, then, capitalism that is preventing the reduction of emissions; it is the good that fossil fuels do for us. The historical adoption of fossil fuels has increased people's quality of life, making it difficult for many to conceive of a good society without them.

A potential set of solutions for these first two problems is to consistently reframe the communication of the climate problem to emphasize how responding to it will not require a loss in quality of life, but actually enhance it. Apprehension or apathetic deprioritization of the climate crisis can thereby be counteracted with an appeal to voters' reasonability and self-interested preference for improving life chances. The reasoning is that electors and the private sector are unwilling to accept solutions to climate change — long a distant and abstract problem — that demand economic sacrifice (Pielke 2010).

In the late 2000s and early 2010s — amid discussion about what the successor agreement to the Kyoto Protocol should look like — prominent thinkers (e.g., Friedman 2009) attempted to address this matter. They argued that if electors and the private sector are unwilling to accept economic sacrifice and are unmoved by talk of the looming ills of climate change, they can be persuaded by appeals to the larger economic and social goods — "collateral benefits" (Gore 2009) — that will occur as a result of pursuing mainstream solutions to the crisis. These include new opportunities to foster economic growth, create jobs, and generate profit. The shift to renewables would also, depending on the country, lessen reliance on fossil fuel energy from regimes under dictatorships or committing human rights violations.

Third, the climate crisis is a public action problem. The Faustian bargain does not only involve individuals' appreciation of the good that fossil fuels do in their lives; it also pits countries' motives against one another. Historically, individual states have had little reason to invest in an expensive transition to untested renewable energies as long as they could see that other states were sticking with fossil fuels; economically speaking, early adopters would see themselves at a competitive disadvantage. But as renewable energy technologies have advanced and achieved cost parity with fossil fuels, there is no longer as much hesitance.

Finally, there are people who simply refuse to accept the truth of climate change (a topic explored in the next chapter on climate change denialism). But this, too, has some straightforward solutions that do not require substantial economic change. Climate change denier propaganda can be debunked and propaganda strategies exposed. The fossil fuel industry's corruption of democracy can be resolved through strategies to elect bold and climate-sincere politicians willing to impose carbon pricing or smart regulations.

THROUGH A CLIMATE JUSTICE LENS

The five questions of climate justice surveyed in the previous chapter raise the moral urgency of realizing truly ambitious programs to achieve emission reductions that will prevent global average temperatures from breaching the Paris climate agreement targets; of not assuming that the existing status quo must be maintained, particularly if doing so means sacrificing those ambitious temperature targets; and of finding support from democratic movements making efforts to represent broad interests, including those of the voiceless and underrepresented. Applied to the neoliberal framework, a climate justice lens emphasizes at least two major concerns: first, the preservation of the neoliberal system over the preservation of a safe climate and, second, a corporate embrace of climate action that signals the crystallization of a mainstream climate politics increasingly dominated by the priorities of major for-profit concentrations of economic power.

System Preserving, but Climate Sacrificing?

Long haunting the neoliberal framework is a need to show that clean energy adoption, capitalist growth, and emissions reductions in line with meeting the Paris Agreement goals are in fact reconcilable. Because the framework rejects restructuring the existing economic order as part of its climate response, the main way to bridge the gap between the need for perpetual economic, market-driven growth and sharp reductions in fossil fuel sources of energy is to turn over major responsibilities to technology. In addition to being clean(er), these technologies must become cheap enough relative to fossil fuels to be widely adoptable, must make available sufficient levels of energy to power the complex of activities that make growth possible, and must offer profit rates that incentivize private market actors to make deep and rapid investment in them. Attempting to bridge the gap in this way has thus always been a gamble that clean tech would eventually progress to the point that all of the above aligns in time to replace fossil fuels before greenhouse gas concentrations commit the world to dangerous climate change. On its own, clean tech has not achieved this goal (proponents would argue fossil fuels have been unfairly subsidized by industry-friendly governments), and so carbon-pricing policy is intended to speed that alignment up. But by and large, intended contributions to global emissions reductions goals announced by individual advanced capitalist countries have tended to be far lower than needed to prevent climate destabilization, showing a stubborn unwillingness to raise carbon prices to a point where emissions might fall rapidly but make growth and profit more turbulent. In other words, where the choice has had to be made, the neoliberal framework has consistently prioritized the preservation of the system over the preservation of a safe climate. Where the world may now need to choose between neoliberal capitalism as usual or a safe climate, this framework seeks to ensure that the only solutions being considered are the ones insisting that these are not mutually exclusive goals. This is a potentially pernicious gamble.

Environmentalism of the Rich

An important new form of environmentalism was becoming prevalent in the 1970s and 1980s. It emerged from the application of holistic, systems-level thinking to global human challenges, and was most prominently exemplified by the interdisciplinary researchers and thought leaders with the Club of Rome, whose famous 1972 report, *The Limits to Growth* (written by Meadows et al.), argued that infinite economic growth cannot be sustained on a finite planet, and that future sustainability would require a steady-state economy (a view explored in greater depth in Chapter 8). At the same time, the environmental movement had succeeded in raising concerns about sustainability to the point that governments were feeling pressured to adopt stronger environmental regulations. Businesses chafed at the thought, fearing that the added costs of compliance would make their products costlier and less competitive, that being forced to go green would send them into the red.

What has come to be known as ecological modernization was a response suggesting that neither of these positions — environmentalists' unnecessarily dour capitulation to natural limits and businesses' chauvinistic dismissal of those limits — would be necessary. Importantly, business could change its ways and not only avoid losing profits but actually find a competitive edge (Hawkin, Lovins, and Lovins 2000; Braungart and McDonough 2002). An important key would be to see nature as a form of capital to be invested in. This *natural capital* approach involves valuing all forms of capital, including the stocks and flows that come from nature, and incorporating that value into policy, planning, and behaviour. In this way, the natural world will be seen as an input into production that ought to be factored into economic calculations, a move towards a more environmental rationality.

A result of attempting to incorporate environmentalism into capitalism is what Canadian political scientist Peter Dauvergne (2016) calls environmentalism of the rich, an increasingly hegemonic form of environmentalism that is stripped of its critical edge, characterized by corporate co-optation of environmental issues — appearing to go green to appeal to more market segments — in order to generate continued profit while failing to confront endless and unsustainable consumption. Its defining feature is "the loss of a 'spirit of outrage' at the underlying structures of exploitation, inequality, and overconsumption that are causing the global sustainability crisis, and by a spirit of compromise with solutions those at the World Economic Forum in Davos can live with" (Dauvergne 2016, 7). Dauvergne argues that the policies and practices composing the environmentalism of the rich get repackaged in various ways to fit the context: as sustainable development by governments, as corporate social responsibility by corporations, as business partnerships and legitimation of market solutions for environmental nongovernmental organizations, and as ecoconsumerism and insignificant lifestyle changes for individuals.

Something of note through the 2010s was the growing assumption on the part of major corporations that they were to play a large role in the climate response. A key moment in the United States occurred in 2009 when the US Chamber of Commerce, described as the largest lobbying group in the country, saw an internal split over its position on climate change, with its leadership opposed to action on the issue, major members like Apple resigning from the group because of this, and a breakaway group, the Chambers for Innovation and Clean Energy, establishing a network of local chambers to demand stronger action on climate change.

In July 2015, the Obama White House launched the American Business Act on Climate Pledge. Several major corporations (Alcoa, Apple, Bank of America, Berkshire Hathaway Energy, Cargill, Coca-Cola, General Motors, Goldman Sachs, Google, Microsoft, PepsiCo, UPS, and Walmart) had signed on to take climate action as part of the lead-up to the Paris climate negotiations. In 2017, hundreds of US firms, including large ones like Nike and Starbucks, urged incoming president Donald Trump not to abandon the Paris climate agreement. Early into the presidency of Joe Biden, in 2021, hundreds of companies urged him to cut the country's emissions in half by 2030. Some of those companies, such as Amazon, Walmart, and Google, are investing massively in powering their operations through renewable energy. By spring 2021, over one hundred multinational corporations had signed on to the Climate

Pledge, a public commitment to report emissions, undertake decarbonization in line with the Paris Agreement goals, and offset any remaining emissions (We Mean Business Coalition 2021).

Billionaires feature prominently among voices offering solutions to climate change. (As a 2021 *Guardian* headline put it, "The latest must-have among US billionaires? A plan to end the climate crisis.") When President Biden hosted a major climate conference in April 2021, Bill Gates, who had recently authored his own book on climate solutions, was a featured speaker.

When major economic power centres and personalities famous for their business success and wealth are elevated to be featured voices on the matter, it is a glaring sign that the " 'spirit of outrage' at the underlying structures of exploitation, inequality, and overconsumption" has indeed been scrubbed away. This approach to environmentalism pushes aside some potentially powerful grassroots solutions, such as massively raising taxes on corporate profits and the wealthy to pay for sweeping decarbonization programs. A critical question at this moment is who gets to decide what to do with the economic surplus generated under capitalism. Why trust major for-profit corporations and billionaires to shape policy and make decisions about where that surplus goes instead of democratic governments and movements?

CONCLUSION

A particular form of liberalism has reigned across the world for the last forty years. As the climate crisis has grown, neoliberal political orders have had to, however reluctantly, make a compromise and permit some modest tinkering with the market in order to address greenhouse gas emissions. For those concerned with climate justice, there is much to be concerned about in an answer to the climate crisis limited to market-oriented solutions, a sign of some neoliberal mass weighing heavily on our political fabric, distorting the nature of policy responses and dragging ambitions downward. As will be seen in a couple chapters, another version of liberalism seeks to remove this mass entirely so that the fuller powers of democratic government can come into play in guiding a kinder, greener capitalism.

We can never know what might have happened if the market-oriented policies we are now seeing adopted throughout the world had been implemented earlier, but it is conceivable they might have been enough to avoid the crisis. Instead, they have arrived late into the game. Part of that is due to the ideological framework itself and its discomfort with government intervention in the capitalist market. But, as the following chapter demonstrates, another part is due to the influence of an ideology that could never make the compromise on this matter that neoliberalism could, that could never come to terms with the idea that there could be anything wrong with a free market and limited government. We turn to that now.

5

RIGHT-WING IDEOLOGY AND CLIMATE CHANGE DENIALISM

"YOU JUST WATCH"

EMILY TOWNSEND HAD HAD enough.

When she, like all other Australia staff at the News Corp media conglomerate, received an email from the executive director about raising funds to help the country deal with its record-breaking 2019–20 inferno, Townsend felt compelled to write the following:

> Thank you for your email regarding fundraising and other support initiatives in relation to the devastating fires.
>
> Unfortunately however, this does not offset the impact News Corp reporting has had over the last few weeks. I have been severely impacted by the coverage of News Corp publications in relation to the fires. In particular the misinformation campaign that has tried to divert attention away from the real issue which is climate change to rather focus on arson (including misrepresenting facts).
>
> I find it unconscionable to continue working for this company, knowing I am contributing to the spread of climate change denial and lies. The reporting I have witnessed in The Australian, The Daily Telegraph and Herald Sun is not only irresponsible,

but dangerous and damaging to our communities and beautiful planet that needs us more than ever to acknowledge the destruction we have caused and start doing something about it. (Samios and Hornery 2020)

That reply — scrubbed from inboxes within the hour, though not before being leaked to the media — went out to all the staff of News Corp Australia, which owns the three dailies mentioned and others that were downplaying the role of climate change in making the blazes so disastrous (Cave 2020; Cranley 2020; Meade 2020; Walton 2020). As the country burned, News Corp outlets blamed the fires on arson or poor forest management — anything plausible sounding, just as long as it was not climate change. Critics noted one particularly revealing absurdity: on the same day that newspapers around the world featured images of the apocalyptic fires on their front pages, the early edition of News Corp's flagship paper *The Australian* featured pictures of picnic races. Before readers reached page 4, where the news about the fires had been relegated, they were treated to an exclusive interview with a "rebel marine scientist" who attacks the science showing that human activity is driving the devastation of the Great Barrier Reef (A. Morton and Smee 2019; Readfearn 2019). Later that month, the fires still burning, News Corp's *Herald Sun* featured a column entitled "Warming is Good for Us" (Rowell 2020).

It's little wonder that Emily Townsend found it unconscionable to work for that sort of company.

In the United States just a few months later, Fox News, another News Corp–owned outlet, similarly de-emphasized the role of climate change in the fires then blazing through the American west coast (MacDonald 2020; Robinson 2020). Tucker Carlson, host of one of the channel's most-viewed shows, asserted to his millions of viewers on September 10, 2020, that, like systemic racism, the climate change that had shifted California's forests into a tinderbox was a liberal invention (Graziosi 2020).

Even the US president got in on things. As the state experienced its worst fire season on record, California government officials hosted Donald Trump on September 14. Among them was Natural Resources Agency Secretary Wade Crowfoot, who urged the president to take seriously the science on climate change and its effect on forests, and to stop, as Trump had been doing, attributing the infernos simply to poor vegetation management.

"It'll start getting cooler. You just watch," Trump responded, baselessly.

"I wish science agreed with you," Crowfoot replied.

Chuckling smugly, the president said, "I don't think science knows, actually," and brought the discussion to a close (Lemire et al. 2020).

All of this occurred within just one year and agonizingly deep into the timeline according to which the world needed to be taking massive action on climate change.

In a rational world, these kinds of things are not supposed to happen. In the face of increasingly destructive impacts, even the staunchest of climate change deniers, we might assume, should see that they were wrong, perhaps even show contrition for their role in preventing action. But that assumes denialism was ever concerned with truth. As an effort to make verifiable or truthful claims about the physical world, the climate change denialist endeavour is an abject failure. The opposition to climate action is not grounded in any body of reputable science that any climate change denier has ever been able to point to. Indeed, one of the more revealing elements of deniers' behaviour is that they never seem to act like they are on the cusp of a major scientific discovery. They are suspiciously incurious about the world-changing implications of having supposedly found major flaws in our understanding of physics and chemistry. They are tellingly uninterested in what it means that an entire scientific field could be a hoax.

That's because the point all along was something different from science and truth. What motivates denialism is not an absence of convincing evidence for climate change, but rather the conviction that climate change cannot be allowed to be seen as an issue serious enough to respond to. If the findings of climate science cannot be reduced to having no status higher than being one opinion among others, then serious changes will have to be made to society. And for some people, that is simply ideologically intolerable.

IDEOLOGICAL SUSCEPTIBILITY TO DENIAL

If you are not used to thinking about the role of ideology, you might note something inexplicable in opinion polls examining attitudes about climate change. You can see it most clearly in the United States, where data goes back far enough to show a persistent, decades-long (Dunlap 2008) split in beliefs about the reality and severity of climate change *along political lines*, with those on the political right being much more likely to doubt that the climate crisis is occurring or is something to be concerned about. Even now — when the impacts of the crisis are more obvious than ever, and there is a well-established consensus among climate scientists that climate change is human driven (Lynas, Houlton, and Perry 2021) — this partisan split persists. In 2022, just 39 percent of US Republican voters believed the effects of climate change had already begun, 33 percent believed that global warming is caused by human activities, and 14 percent believed it would pose a serious threat in their lifetimes, compared to 85, 91, and 73 percent of Democrat voters, respectively (Gallup 2022).

This pattern is not just reserved for the United States. In May 2021, just 34 percent of Canadians who voted for the Conservative party in the 2019 federal election correctly stated that climate change is a fact and is mostly caused by human activity (compared to an average of 70 percent for Canadians as a whole and around 90 percent of voters for all other major parties). Of those Conservative voters, 43 percent believed climate change is a fact but is mostly caused by natural changes and cycles, while 17 percent of them believed that climate change is a theory that has not been proven (Angus Reid Institute 2021). Similar large partisan splits have been noted in Australia, the United Kingdom, Germany, the Netherlands, and Sweden (Funk et al. 2020). This effect has been observed not just in polls but also in experiments and across studies (Kahan, Jenkins-Smith, and Braman 2011; Hornsey et al. 2016). And, as indicated in the introduction of this chapter, it is media and political leaders on the political right alone who are pushing denial in mainstream discourse.

What all of this suggests is that there is something in right-wing ideology motivating many of those holding it to reject the science that demonstrates climate change is happening, is driven by human-caused greenhouse gas emissions, and is serious. To understand why that occurs, we need to understand this ideology and what about it creates an inclination towards denial.

Let's begin with two defining and overriding right-wing concerns: that all individuals have strong and equal rights to liberty and holding property. First, a good society, in this worldview, must maximize individual freedoms as much as possible. Freedom, of course, is a contested concept and can be defined in more narrow or expansive ways. In the narrower senses, freedom is understood as the range of choices that people can make free from coercion — that is, from credible threats of force. We are not all that free, after all, if our choices can be directed, manipulated, or circumscribed out of fear that choosing incorrectly will be met with destruction of our means of meeting basic needs, corporal punishment, or even death. More expansive senses of freedom, on the other hand, have to do with the range of options we actually have the means to achieve. These are concerned with minimizing not just coercion but also the material deprivation that occurs throughout a society. The first, narrower conception of freedom is typical of right-wing ideology and is often captured, more simply, with the term *liberty*.

Second, a good society, according to this worldview, is also one that recognizes the special status of the right to our possessions — to the goods we have purchased or been given, the money we earned or inherited, the lands or buildings we hold title to, the goods and even ideas (or intellectual property) we have originated. People can and will thrive only insofar as they can be sure that the possessions they have rightfully acquired will not be seized or cheated from them. If there is no guarantee of rights to possession — if someone can take for themselves what another person has earned, produced, or invented — what incentive does anyone have to provide effort, labour, ingenuity, or time to productive activities?

These two foundational concepts lead to a particular vision of the state and the economy that is essential for understanding why denialism has taken root on the political right. As we will see, this vision has extreme difficulty accommodating the kinds of social changes necessary to respond to climate change.

Let's start with the vision of the state. The right's overriding concern for liberty and property make the ideal state a highly restrained one. The more a state expands its role, the more scope it has to coerce people, rendering a society less free. It should not, therefore, be an activist state attempting to promote social aims. New laws and regulations to promote, say, greater equality, tolerance, public health and safety, or sustainability make previously permissible behaviours suddenly punishable. (One example that occurred during the writing of this book was the right-wing opposition, galvanized by calls for "freedom," against government-imposed lockdowns and mask and vaccine mandates to mitigate the COVID-19 crisis. In Canada, this opposition culminated in a self-styled "Freedom Convoy" of truckers occupying the streets of Ottawa for close to a month in early 2022 to protest further restrictions.) Meanwhile, new or expanded programs to promote social welfare raise government spending, demanding a rise in taxation — and what is taxation if not the state forcing individuals to part with some portion of their wealth to fund its goals? Rather, the ideal state should stick, first, to protecting individuals from coercion by each other and outside parties (e.g., invading states). Second, it should enshrine strong laws that protect individuals' rights to their property, laws that include the upholding of contracts. The latter are essential for determining who can possess which rewards from cooperative economic ventures; we have little reason to work with or for someone if, at the end, they renege on their promises and claim everything for themselves. Government spending should be restricted to only those functions essential to society that are unlikely to work well if left to nonstate actors, typically law enforcement (e.g., policing, courts), military, fire services, and some infrastructure services (e.g., roads, airports, ports, sewers).

This political vision of the restrained state is complemented by the economic vision of an unrestrained capitalism. Only in a free market can individuals be at their fullest liberty to make the best decisions for themselves with their own abilities, wealth, and property. That alone is reason to let the market do as it will. But what is more, this same free market naturally creates a meritocratic socioeconomic order where everyone gets what they deserve according to the wisdom of their economic choices. Spending impulsively or beyond one's means, being unwilling to work, performing poorly at one's job, making unwise investments, failing to save for retirement — all of these are punished thanks to the market, which provides nothing to anyone lacking the means to acquire it. Without these consequences, individuals have little reason to act differently, particularly if they can expect some form of public bailout or assistance, and it would be unfair to force others to save these people from having made bad decisions. (This is a key reason welfare-state programs tend to be viewed with enormous suspicion by the right.) And the market does not just punish. It rewards, often lavishly, the individuals who provide economic value to a society, whether in the form of dedicated and hard work, specialized abilities or expertise, investments in productive ventures, or innovative, superior, new technologies, prod-

ucts, processes, or services. Hierarchies based on socioeconomic inequalities are not only natural and inevitable then, but even desirable.
They result from rewarding the tendencies and forward-thinking insights that, though carried out in self-interest, benefit the wider society.
It's this vision of a free, meritocratic economy that is constantly threatened by those calling for system change in the name of justice and equality — and now sustainability — and proposing a variety of solutions,
whether in the form of increased economic regulations, higher taxes, or
redistributive policies, all of which would reduce individual liberties by
expanding the coercive role of government.

The above is a fairly classical depiction of the right-wing worldview,
drawn broadly enough to encompass the vision of a good society that,
whatever else they may disagree on, both a conservative and libertarian
would want to see achieved. A good society for those on the political
right is one guaranteeing individuals an expansive degree of liberty to live
life as they want using whatever means they are able to lawfully acquire.

For people holding intensely to this belief system, climate change
poses a seismic threat — not because of what harms it will inflict on
the physical world but because it shakes the foundations of this worldview. If climate change is the real and serious threat the science says it
is, then it is absolutely necessary to respond to it. But there are no responses that are fully compatible with the ideology *and* at the same time
capable of reducing greenhouse gas emissions at anything like the rate
required. The types of policies put forward to address the climate crisis
all violate core right-wing beliefs. People *cannot* be free to make whatever decisions they want in the market about how much fossil fuels they
want to use in perpetuity. Government *does* have to intervene in major
portions of the economy — transport, heat and electricity generation,
agriculture, buildings — with new regulations and taxes, and possibly
even large public investment projects. If we shape society according to
a rigidly right-wing ideology in the face of the climate crisis — if we let
individuals use whatever forms of energy they want, reject a role for
government in the economy, and simply trust to the wisdom of the free
market — then so do we destroy the ecological basis for the contemporary way of life and even human existence. It is hard to think of a worse
condemnation of a belief system.

And it's here where the potential for denialism lies. Those on the
right face a difficult dilemma: either accept reality and permit some

part of their belief system to be violated or find some way to question the reality undermining the ideology. Denialism offers a psychological defence mechanism that soothes people's minds, just at the cost of a habitable climate.

But the spread of denialism was neither automatic nor necessarily natural. Something was needed to complete the link from an ideological susceptibility to denial to a full-blown embrace of it.

THE DENIAL MACHINE

Climate change denial has been elicited, organized, and perpetuated with the singular goal of preventing climate policy for as long as possible. Researchers of climate change denial (Dunlap and McCright 2011; Treen, Williams, and O'Neill 2020) characterize what generates and sustains this state of misbelief as a "denial machine." It is an apt description for a system composed of interlocking parts each carrying specialized functions, whether that be funding denial, generating denier myths, or propagating them.

Financers of Misinformation

The first part of the denial machine consists of corporate and philanthropic actors whose main role over the years has been to initiate and then fund the denial project. The dominant corporate actors have been fossil fuel companies, who understood from the start that their profitability would be undermined by any serious effort to address climate change. One of the more egregious examples was revealed in 2015 through investigative reporting by *InsideClimate News* and the *LA Times*. The exposé highlighted the early role that Exxon played in conducting legitimate climate science research that showed human-caused greenhouse gas emissions would lead to significant climate change, only for the corporation to then become an originator and leading financer of the climate change denial project (Cook et al. 2019; Supran and Oreskes 2020). Later reporting showed how the company continues to obstruct climate action into the 2020s (Carter 2021).

The philanthropic actors, meanwhile, consist mainly of politically conservative foundations looking to fund efforts to protect or advance right-wing political projects, particularly those aimed at reducing taxation and environmental regulation. In a few notable cases, the philanthropic actors are themselves closely aligned with the fossil fuel industry. The Koch Foundation, funded by the Koch fossil fuel empire, is the preeminent example. Between 1997 and 2018, the foundation gave more than US$145 million directly to groups denying climate change (Greenpeace n.d.).

Producers of Misinformation

Because there is no scientific case for doubting that climate change is happening, being driven by human activities, and has serious consequences, deniers had to manufacture opposing arguments that could seem plausible to audiences. The playbook for doing so had already been perfected in the war on the science linking tobacco use to disease. As the tobacco industry faced increasing evidence that its product was lethal, it financed a propaganda campaign spearheaded by public relations gurus and scientists holding an ideological aversion to government regulation who were willing to lend credence to the denial project, whom Oreskes and Conway (2010) named "the merchants of doubt." The same play was used to delay action on the problem of acid rain and ozone depletion. In each case, the doubt merchants applied their major realization: that you can still win the political battle even if you are losing the scientific one. All you have to do is manufacture enough doubt in that science.

This is where right-wing or free-market think tanks come in. Like other think tanks, they produce a variety of research through discussion papers, reports, blog posts, and more, which are intended to shape policy discussions on an array of issues, in this case to push for policy changes that would shrink government, reduce public spending, deregulate industries, lower taxes on corporations and the rich, and so forth. But precisely because these are the policies that cannot be pursued if we are to respond to climate change, these same think tanks have tended to be some of the most prolific sources of climate change denier arguments. In the United States, right-wing think tanks with a history of climate denialism include the Heartland Institute, the Competitive Enterprise Institute, the Committee for a Constructive Tomorrow, and the Cato Institute (cofounded by Charles Koch). In Canada, we find the Frontier Centre for Public Policy and the Fraser Institute. Australia has its Institute of Public Affairs. Alongside think tanks are a variety of advocacy or front groups more focused on the climate change issue than on general right-wing policies (Union of Concerned Scientists 2017).

These think tanks and front groups engage in "information laundering" (Shulman, Abend, and Meyer 2007). The public would be highly suspicious of reports from the fossil fuel industry that just happened to find that its products were not causing dangerous climate change. And so, in the same way that an organized crime syndicate will invest its revenue in legal operations to prevent raising suspicions about the sources of its wealth, the fossil fuel industry has bankrolled outside groups that push convenient myths on climate change.

The role of contrarian scientists should be mentioned here. There exists a small coterie of scientists who hold credentials in relevant fields that are leveraged to lend the denial endeavour more credibility. These contrarians frequently associate with think thanks to lend themselves an air of legitimacy. For instance, prominent denier Patrick J. Michaels could for years make his frequent media and public appearances under the impressive-sounding title of "Director of the Center for the Study of Science at the Cato Institute" instead of the more accurate "long-discredited man about to spew nonsense."

Propagators of Misinformation

To be effective, the talking points generated by the producers of misinformation need to be transmitted and disseminated to laypeople (particularly those ideologically receptive to the messaging), most of whom are unlikely to read through obscure think tank reports or attend denier conferences.

Echo Chamber 1: Influencers

Enter the influencers, figures with both a talent for communicating simple but impactful and memorable messages to mass audiences and a platform from which to do so.

The first key member of the influencer echo chamber is news media. In the US television network landscape, Fox News has long played a unique role in pushing denier talking points into the American mainstream (Theel, Fitzsimmons, and Greenberg 2012). Studies have found that most of the channel's climate coverage contains misleading portrayals of the science and resorts to fearmongering when discussing proposed solutions (Huertas and Kriegsman 2014; Public Citizen 2019). Compared to those who watch other major television network news media, Fox viewers are consistently the most deeply misinformed and unconcerned about climate change (Gustafson et al. 2020).

Denial also comes from Fox News's competitors even further to the right in the US cable landscape, like Newsmax TV and One America News Network. In Australia, the right-leaning television channel SkyNews (like Fox, owned by News Corp) leads the denial charge. A 2020 study found that more than a third of the channel's audience were unconcerned about climate change, the highest figure (tying with audiences of AM radio and Fox News) of any news source. In 2021, GB News launched in the United Kingdom, and wasted little time before platforming deniers (de Ferrer 2021).

In print media, columns and op-eds in right-leaning outlets have long been reliable sources of climate change denial. Columnists can rapidly produce large quantities of material and can do so in response to important events relating to climate politics such as negotiations for a new international climate agreement, imminent prospects of regional or national climate legislation, the publication of a major scientific report, or the rise to prominence of a public climate figure like Al Gore or Greta Thunberg. In the United States, denier columns and op-eds feature frequently in the *Washington Times*, *Wall Street Journal*, and the *New York Post*, the latter two owned by News Corp. But nowhere does News Corp extend more widely over a media landscape than in its home country of Australia, where the conglomerate owns just over half of the newspaper market (Evershed 2020). Even as late as 2019–20, 45 percent of the content on climate change across just four Australian News Corp publications rejected or cast doubt on the findings in the scientific consensus (Bacon and Jegan 2020). The conglomerate's influence is also pronounced in the United Kingdom, where it owns *The Times* (and *Sunday Times*) and *The Sun*. Canada's most prominent examples are found in the more right-leaning outlets of the Postmedia empire, including the Sun Media chain and the *National Post*. In recent years, there has been a spurt of influential internet-based news media with ultra-conservative or far-right slants regularly spreading climate misinformation. In the United States, these include The Daily Wire, The Daily Caller, Breitbart Media, and Blaze TV.

A second set of key actors in the influencer echo chamber have been denier bloggers, the most prominent being found at *Watts Up With*

That?, *Climate Depot*, *Junk Science*, and *Climate Audit*. In producing reams of think pieces and timely reactions to climate news, they play a role similar to that of right-wing columnists. But not having to appeal to wide audiences, these websites can specialize in denialism, conducting deep dives into the minutiae of denier arguments, unrestricted by editorial oversight.

The last set of actors in the influencer echo chamber are politicians who dutifully parrot denier talking points, typically after receiving significant campaign financing from the fossil fuel industry. No party in the advanced capitalist world has been as deeply saturated with denialism as the US Republican Party, a tradition that goes back to the George H.W. Bush years (S. Waldman and Hulac 2018). The Center for American Progress Action Fund has tracked the number of climate change deniers in the US federal legislature through the 112th to the 117th Congress. From 2011 to 2021, 25–30 percent of representatives in the House and 30–38 percent of senators have been climate change deniers, all of them Republicans (B. Johnson 2010; Spross, Germain, and Koronowski 2013; Ellingboe, Germain, and Kroh 2015; Moser and Koronowski 2017; Hardin and Moser 2019; Drennen and Harden 2021). Together receiving millions in election campaign contributions from the fossil fuel industry, they have long succeeded in blocking significant climate legislation in the country and provided examples (or *elite cues*) to the party's rank-and-file voters on how to see climate change in ideologically acceptable ways.

Echo Chamber 2: The Public

The final key role in the denial machine has been handed over to the public. The body of arguments that deniers have produced creates a deep rabbit hole into which people motivated to hear out the "other side" of the climate debate can fall. Laypeople can feel as though they really are conducting their own research, immersing themselves in some secret and forbidden body of truth that in actuality is nothing more than a midden of disinformation excreted by the charlatans and propagandists serving the denial machine. In the era of social media, ordinary people face no barriers to recirculating denier arguments that can reach wide audiences on Facebook, Twitter, 4Chan, or subreddits like r/climateskeptics. The use of memes in online communication provides a high-efficiency means of presenting and propagating emotionally poignant but information- and context-stripped messages easily applied to climate change (Wilson 2020).

TYPES OF DENIAL AND THE LOGIC OF THE ILLOGICAL

That denial machine has been prolific in conjuring and disseminating several lines of denial (Coan et al. 2021). *Trend denial* describes those attempts to show that there is no significant long-term warming trend in the earth system. *Attribution denial* acknowledges warming but claims it is caused not by human activities but by something else. *Impact denial* casts doubt on the severity of climate change. A sophisticated variant comes in the form of "lukewarmers," figures presenting as concerned environmentalists who even acknowledge that climate change is human caused only to then oppose most actions to address it while touting the "benefits" of continued heavy fossil fuel use. What might be called *proponent-reliability denial* alleges that climate scientists, mainstream media outlets, and activists are spreading alarmist misinformation driven by unsound scientific methods, political bias, or ulterior motives while censoring dissenting scientific views concerning whether human-driven climate change is happening. This line of denial benefits from conspiracy theories purporting, for example, that scientists are in on it for research grants or that climate change is a Trojan horse for authoritarian left-wing ideologues to steadily expand state control over people's lives. The last — and now potentially most dominant (Coan et al. 2021) — line of denial is *solutions denial*, which attempts to derail support for ambitious climate policies by denying that transformative changes are necessary, desirable, or workable (Lamb et al. 2020; Mann 2021, chaps. 6–7). It is particularly insidious today because its arguments, though misleading, can evade the stigma in mainstream political opinion against outright climate science denial while preying on legitimate concerns about fair and effective climate policy to undermine action in the last years left to keep warming well below 2°C.

Notice something strange about these lines of attack: they contradict one another. What is a person to believe — that the earth is not warming *or* that it is warming but not because of greenhouse gases *or* that it is warming due to greenhouse gases but things will be fine *or* that it could be bad but we cannot and need not do anything about it? Simultaneously proposing all these possibilities makes the denier body of arguments rife with incongruity. Overzealous deniers even contradict themselves in the same work (Lewandowsky, Cook, and Lloyd 2016, 186–87). As a means of producing a consistent and plausible view of the world that can rival a dominant scientific paradigm, adopting this overly protean and self-contradictory stance is utterly irrational — observers have derisively called it the " 'Alice in Wonderland' mechanics" (Lewandowsky, Cook, and Lloyd 2016) and "the quantum theory" (Cook 2017) of climate change denial — and it reveals that coherency was never the function of the denial machine. Its true function was to perpetuate enough doubt to prevent action, to hurl fistfuls of sand into the political gears working to produce a climate response. The arguments that are admitted into the denier canon have little to do with robust evidence or compatibility with other denier arguments, and are instead about the degree to which they can create doubt. The only consensus this canon has to reflect is that the threat of human-driven climate change must not be taken seriously. What is utterly irrational for the purpose of investigating and understanding the world turns out to be fully rational as a means of maximizing the number of arguments that deniers can turn to. Being able to shift between different denial states as needed grants those engaged in the denier project enormous flexibility in choosing how to contest just about any part of the scientific case for human-caused climate change depending on the context. No one should expect committed deniers to stop denying regardless of how obvious climate impacts become. Ideological safeguarding, not truth, was always the point.

But understanding the different forms that denialism can take leaves us still unprepared to understand the many techniques deniers can use to perpetuate their arguments. Cook (2020) has assembled these under a taxonomy he refers to by the acronym FLICC, with each letter representing an entire scheming family of denier techniques: fake experts, logical fallacies, impossible expectations, cherry-picking, and conspiracy theories. It's a reminder there is no shortage of tactics that deniers can use, and the denial machine will never really run out of arguments.

So what is to be done? The years of the Donald Trump presidency were powerful demonstrations of what occurs when deniers come to power. Preventing that sort of fossil fuel industry–allied politician from ascending to power is an important matter of climate justice. As Chapter 10 covers, through the 2010s the climate movement pivoted to direct much of its energies to eroding the political strength and reputation of the fossil fuel industry, the historical bankroller of denier propaganda and politicians. The most immediate climate justice problem here is that as long as the denial machine persists in validating denial, its workings will continue obstructing or slowing down climate action in these crucial years. How can the link between the machine and susceptible populations be broken?

THROUGH A CLIMATE JUSTICE LENS

For one morning, using just a few trucks and some bamboo scaffolding, UK members of the nonviolent, direct action climate movement group Extinction Rebellion blocked the road leading from printing presses publishing newspapers with a history of climate change denialism.

In response, Conservative prime minister Boris Johnson tweeted, "A free press is vital in holding the government and other powerful institutions to account on issues critical for the future of our country, including the fight against climate change. It is completely unacceptable to seek to limit the public's access to news in this way." The centre-left Labour Party was similarly unimpressed, noting, "A free press is vital for our democracy. People have the right to read the newspapers they want. Stopping them from being distributed and printers from doing their jobs is wrong." Condemnation came from elsewhere as well. Predictably, the right-wing media hated it, as evinced by an astounding *Daily Mail* headline: "The middle-class eco rabble who want to kill off free speech: Extinction Rebellion activists moan their climate change doomsday message isn't being printed on newspaper front pages EVERY DAY... as they block access to national presses." But even a spokesperson for the left-leaning Guardian media company denounced it (Iqbal 2020).

The tactic of targeting media outlets that push denier propaganda poses some challenging questions for a response based on climate justice. Let's consider those reactions characterizing it as an attack on free speech. One question in response is whether there exist messages of a nature too destructive to tolerate in a free society and whether, therefore, there ought to be some form of limits on those forms of speech. If climate denier propaganda is considered protected speech, it creates an absurdity: laws protect disinformation that contributes to the obstruction of efforts to protect society from widespread — and possibly cataclysmic — disaster. Climate disinformation would not be significantly different then from propaganda being spread by domestic allies of an invading enemy army in wartime. To do as those voices did above — to simply condemn Extinction Rebellion's action on liberal grounds highlighting the importance of free speech for maintaining an open society and accountable government — gives us no real answer as to what to do in those instances when free speech has been so exploited as to become a threat to society.

But what about climate change deniers spinning their deplatforming to their advantage, twisting it through bad-faith arguments to cast themselves as victims of the authoritarianism so many of them baselessly claim is the climate movement's true end goal? The kind of statements they would issue are predictable: "Big government wants to control what kind of car you can drive, how much meat you eat, *and now even what you can say!*" But deniers already cry authoritarian wolf about every serious climate change response, meaning there is no need to hold back on a response out of fears about how they will twist it. Another argument, more difficult to dismiss, is that legitimizing deplatforming actions unsheathes a double-edged sword. Consider the debate over what kind of economic system we might want, a matter discussed throughout this book. How much trust should we have that extending acceptance of restrictions on speech would not lead to anticapitalist speech becoming censored (or antiwar speech being shut down in times of conflict)?

But let's look at why Extinction Rebellion actually carried out the action in the first place (and why American members carried out a similar action in New York in April 2022). To them, characterizing the UK press as "free" misrepresents a media landscape dominated and heavily shaped by highly concentrated, privately owned, for-profit conglomerations. Indeed, it was this problem — how civil society should respond to a corporate media oligopoly that has the capacity to ignore climate change or deliberately saturate the conversation on climate with misinformation — that the road blockade was intended to draw attention to rather than to the matter of whether media outlets should be free to publish denier propaganda (Extinction Rebellion 2020a).

If there were to be some measures for preventing deniers from polluting the public conversation on climate change, what form might they take? No one serious is calling for the state to punish individual climate deniers with fines or incarceration for expressing their views in news media. Instead, there are calls for restricting the ease with which they can exploit major media platforms with wide reach in order to spread disinformation.

A corporate press is vulnerable to some tactics that could be of use. The political economy of for-profit news media is such that outlets must be able to attract advertisers (or to put it another way, viewers are the product that news media sells to advertisers). As corporations grow more concerned with showing efforts to shrink their carbon footprints, there is an opportunity for movements to pressure them to take that into their advertising policy and deny ad revenue to major media sources that continue to give space to climate change deniers. If the corporations are hoping to project an image of sustainability, they ought to refrain from financially supporting outlets that would reverse those efforts. In 2019, the left-leaning US media watchdog organization Media Matters launched the "UnFox My Cable Box" campaign in hopes of rallying a response against another major source of Fox News's funding. Because the channel is included in a number of cable and satellite provider TV bundles, subscribers end up paying an estimated US$1.8 billion annually in carriage fees to Fox News whether or not they have any interest in watching the channel.

Another important arena of contestation is social networking sites, where climate misinformation is rife (Avaaz 2020). Being privately owned, social media companies are under no obligation to protect denialism as a matter of free speech, and so advocating for better content moderation bypasses that fraught matter. There have been ongoing campaigns to pressure Facebook, YouTube, and Twitter to moderate the extent to which misinformation on climate change (among other issues) can be spread freely. The record so far is unimpressive. The case of Facebook has been illustrative. In 2020, the social media giant established a Climate Science Information Center (later renamed Climate Science Center) meant to act as a resource for people looking to understand more about the science of climate change, though it is hard to see how this addresses the political ideological drivers of denialism. The climate solutions it presented were, furthermore, anodyne, asking visitors to turn off lights and unplug electronics when not using them, recycle, buy local foods, and drive less. Facebook also instituted a fact-checking program to flag climate disinformation, but it was shown to be vulnerable to loopholes (Atkin 2020). At the same time, in just the first half of 2020, advertisements on Facebook promoting climate change denial were viewed eight million times (InfluenceMap 2020). A 2021 report from Stop Funding Heat estimated that users' views of climate misinformation posts on Facebook were far higher than visits to its Climate Science Center. Twitter, meanwhile, has its own loopholes through which oil company ads can spread misinformation (Atkin 2021b).

In the aftermath of the events of January 6, 2021, when far-right Donald Trump supporters carried out an insurrection at the Washington, DC, Capitol building in an attempt to overturn federal election results, social media companies took a more aggressive stance on disinformation. Animating the violent riot was a belief, as fanatical as it was baseless, that the November 2020 election had been stolen by Joe Biden. It found feverish support among Trumpists and conspiracy theorists imbibing and then recirculating internet misinformation. The events occurred in the context of years of violent right-wing extremism being cultivated in unmoderated "free speech" online communities, contributing in the worst instances to spree killings undertaken in the cause of white nationalism. Climate change denialism has no more credibility than any of the major conspiracy theories that social media companies began cracking down on in the early 2020s — whether about the 2020 US election being fraudulent or COVID-19 being a hoax — and, to the extent that denialism prevents action to avert the crisis, is far more dangerous.

COMING TO TERMS WITH CLIMATE CHANGE

With the dominant economic paradigm having shifted rightward in neoliberal times, there has been reduced policy space through which standard right-wing parties could distinguish themselves from liberal parties and still be seen to take the climate crisis seriously. Those on the right who do acknowledge the threat of climate change have tended to propose carbon-pricing solutions (Schwartz 2017) that neoliberal parties occupying the centre of many countries' politics have already begun implementing. This should always have been the right's primary response, but the long-unwavering flirtation with denialism meant they were long ago outmanoeuvred by more centrist neoliberals. There are potential signs of change. For example, in spring 2021, Canada's federal Conservative party, which had long ignored climate change, put forward a version of carbon-pricing policy. Another set of solutions those on the right have begun discussing (Roberts 2021b) fall under geoengineering, a series of direct interventions in the earth's climate system to address climate change, which is discussed in the next chapter.

But ideologies also have malevolent potentials that awaken when their traditional beliefs can no longer explain the world, as recent years have seen on the right. For those believing that the pathway to success is hard work and personal responsibility in a society of deregulated markets and small government, the wage stagnation, debt, and precarity that workers experience in neoliberal times — despite a lifetime of hard (over)work — is incomprehensible. Right-wing parties certainly cannot turn to the left to offer a positive, restorative economic vision. What they and their allied media can do though is feed a narrative that, however unsubstantiated, *feels* true: that all that stagnation is due to progressive governments permitting (or even encouraging) "others" — minorities, immigrants, refugees, women — to "cut in line," pushing the truly deserving hard workers further and further behind in the grand, slow-moving queue for society's rewards (Hochschild 2016). This narrative helps explain the rise through the 2010s in nationalist anti-immigrant and antirefugee sentiment throughout the Western world, as in Europe's identitarian movement and, most prominently, in the ascent of Trumpism in the United States.

The increase in authoritarian longings in the United States (MacWilliams 2020) and elsewhere are also made legible through this narrative: they are yearnings for a political strongman to put the wrong people back in their place and restore the rightful to theirs. It is a situation primed for opportunistic demagogues to mobilize political support by tapping into that line-cutter narrative — fostering a sense of bitterness, victimhood, humiliation, status loss, and disaffection with traditional elites and their institutions — and promising to do something about the undeserving "them" who have taken the opportunities that should belong to "us." Status paranoia can pull violent extremist belief from an isolated fringe to the mainstream. In research conducted in the aftermath of the 2021 US Capitol assault, Pape (2022) estimates that 8 percent of Americans, or twenty-one million people, believe both that Joe Biden stole the 2020 presidential election and that the use of force is justified in restoring Donald Trump's presidency. These defining beliefs of what researchers have dubbed America's "insurrectionist movement" appear to be motivated by a still deeper and stronger driver — its members' widely shared subscription to the "great replacement" conspiracy theory, which claims that liberal politicians are using nonwhite immigration to remake the country's demographics and reduce the political power of whites. If that is its basis, the insurrectionist movement could see even further growth. A study found that 32 percent of US adults believe there is a group of people using immigration to replace native-born Americans for political gains (AP-NORC 2022). In these "post-truth" times, when objective facts count for less in political arguments than manipulative emotional appeals nigh impervious to debunking, when proof is immaterial in convincing people that those like "us" are threatened from some dangerous "them," we would do well to heed the frequent warning of holocaust historian Timothy Snyder (2017, 65–71; 2021): "Post-truth is pre-facism."

In this same context, recent years have seen a strong strain of antileftism or antiprogressivism on the right. Political commentators such as Steven Crowder, Jordan Peterson, Dave Rubin, and Ben Shapiro attracted large groups of often younger and impassioned followers by offering antileftist takes rejecting political correctness (or "wokeism"), socialism, and concerns about systemic racism, sexism, and inequality. Fellow travellers on the more extreme alt-right echoed these sentiments but added a more intolerant twist, espousing views that were openly Western chauvinist, white nationalist, Islamophobic, or transphobic. In a period when the right-wing political project could offer little in terms of a positive social vision, this antileftism offered a way to stimulate continued political fervour and vitality; in the absence of a vision to work towards, it offered an enemy to work against. American commentators (e.g., Serwer 2018; Beauchamp 2019; P. Waldman 2020) observed how the country's right was primarily being animated not by traditional ideological views but by partisan sadism and malice directed at hated outgroups.

With respect to climate, these more sinister ideological potentials that have emerged on the right can open political space for econationalism or ecofascism, movements that tie together a project of environmental concern with nationalism or, in the direst cases, racial supremacy. They hold, in essence, that the purity of nature is to be preserved only for members of the deserving nation or race. As a case in point, Fox News personality Tucker Carlson, who is widely criticized for mainstreaming white nationalist talking points (Confessore 2022; Ramírez 2022; Yourish et al. 2022), including about the "great replacement," said the following in a 2021 interview: "Unrestrained mass immigration has also put a huge strain on the natural world, just the amount of pollution and litter and destruction that's generated by the movement of hundreds of thousands of people, unrestrained, across the border.... I've never heard anybody mention that. And it's so awful." Econationalism might allow a radicalizing right to accept climate science because of the way it could be manipulated to justify anti-immigrant sentiment, particularly given the potential of climate change–impelled migration and displacement, and to antagonize progressives and leftists concerned with racial and migrant justice. Disturbingly, there are signs of this already (Aronoff 2019; Klein 2020, 40–49; Kaufman 2021; Malm and the Zetkin Collective 2021).

CONCLUSION

We can never know for sure what might have happened in the climate struggle if the deniers had not been able to construct such an influential machine that politicized climate change the way it did on the political right. But it is reasonable to assume that in a world where Koch Industries and ExxonMobil had not funded think tanks conjuring climate denial arguments, where Rupert Murdoch had not been free to control a media empire that, unconscionably, pushed denial to millions for a generation, where Donald Trump did not have four years in power to delay climate action in the crucial post–Paris Agreement period, we would be in a much better place. There might never have been a world without climate change deniers, but there might have been one where they were isolated on the fringe, unable to wield political influence to slow the climate response over the last generation. That influence has come at a cost to so many around the globe, who have to face the consequences of a more dangerous climate. There is no real limit on the number of people that deniers would allow to suffer for their ideological project. Morally, the chief operators of the denial machine are engaged in one of history's most profoundly and unspeakably despicable endeavours.

The rank and file of the ideological right has some serious soul-searching to do. To reject science, as so many of them have done with climate change, is deeply irrational. Without legitimate means to identify and reject disinformation, they are highly exposed and vulnerable to the distortions of propagandists, charlatans, conspiracy theorists, overconfident amateurs, disreputable contrarian scientists, the duped and misinformed, and so on. Denier views are immediately discrediting to any person who professes them, betraying a glaring failure — even inability — to properly evaluate the validity of evidence for competing arguments; if someone is this egregiously wrong about a major political issue, what else are they wrong about? Whatever show lay-deniers make of being independent thinkers asking critical questions and doing their own research, the one thing they most clearly communicate is that they were gullible, easily tricked and brainwashed by the denial machine's ideologically comforting propaganda. It ought to be deeply embarrassing and scandalous to conservatives and libertarians that something as baseless and easily debunked as climate change denialism was able to burrow so deeply into right-wing politics. It should also be a wakeup call. In the end, for all their worries about being controlled by "socialist" big government, those on the right who bought into denialism were all too easily manipulated by disinformation from their supposedly allied media, think tanks, and politicians.

6

GEOENGINEERING

PERHAPS THERE IS A way of understanding and addressing the problem of climate change that never has to involve much more than its physics. The climate will continue to change until the amount of energy the earth emits out into space is in balance with the amount of energy absorbed from the sun. All we really need to do then is alter either or both sides of the incoming–outgoing energy flow. All we need to do, in other words, is change the planet. Geoengineering is the name given to a series of direct and typically large-scale interventions in the earth's climate system intended to address human-caused climate change. It tends to be divided into two kinds.

What are commonly called solar-radiation management (SRM) technologies take on the incoming side of the planet's energy flux with the aim of lowering the amount of solar energy absorbed by the earth. These technologies turn primarily to enhancing the planet's albedo, the effect by which brighter-coloured surfaces reflect more light back into space than do darker surfaces. For instance, stratospheric aerosol injection would see planes or air balloons deployed to release chemicals (e.g., sulphur dioxide, sulphuric acid, or hydrogen sulphide) into the lower stratosphere in order to introduce or encourage the formation of aerosol particles, emulating the global cooling effect created by volcanic eruption plumes. Ships, crewed or automated, can be fitted with tall masts that spray salt water skyward and spur the generation of bright clouds ("marine sky brightening"), or they might stimulate water to create bubbles that brighten the ocean surface, like the white wake behind a fast-moving vessel. Some SRM ideas stretch out to the far reaches of science fiction, like designs to launch fleets of mirrors or shades into earth's orbit to block a portion of the sun's light, or plans to bioengineer crops to be brighter. (Cloud thinning, another approach sometimes lumped in with the above, would reduce the thickness and longevity of cirrus clouds, which on average have a net warming effect on the earth. Because the approach does not deal with solar radiation, it has been suggested that the term *radiative forcing geoengineering* is more accurate to account for the above family of techniques [Lawrence et al. 2018, 3].)

Carbon dioxide removal (CDR) technologies take on the outgoing side of the planet's energy flux, aiming to withdraw from the atmosphere the greenhouse gases frustrating the journey of space-bound infrared energy. As with SRM, there are a variety of proposals for CDR technologies. One approach is *afforestation*; to cultivate forests where there are none or to make existing forests denser involves encouraging the growth of trees, which take carbon from the atmosphere as part of their formation. *Biochar* involves heating biomass in environments without oxygen (a process called pyrolysis) to leave behind stable forms of carbon that can be mixed into agricultural soils. *Ocean iron fertilization* would introduce iron particles into the ocean to stimulate the growth of phytoplankton, which take carbon from the atmosphere and carry it upon death to the ocean bottom. *Weathering* is a process where silicate rocks are crushed and dispersed over a wide area to take up carbon dioxide.

But the last CDR approaches, what we might call the "capturing" kind, have probably gained the most attention. *Bioenergy with carbon capture and storage* involves growing organic matter (which takes carbon from the atmosphere), burning it for energy, and capturing and then burying the emissions. *Direct air capture* (DAC), meanwhile, involves the scaling up of machines that take carbon out of the ambient air. Some years ago, a "carbon-sucking wall" gained mainstream notice, with popular articles featuring an artist's rendition of a massive structure composed of devices capable of scrubbing the air of greenhouse gases.

Another family of responses does not fall into the usual divide of SRM or CDR categories and remains, for now, on the fringe. For lack of a better term, this class of "nullifying" technologies would prevent the occurrence of weather extremes caused by climate change or terminate them once in effect. One can read of high-powered lasers that might one day be capable of eliminating the storm formations that culminate in torrential downpours or triggering the creation of rain clouds over regions afflicted with drought (Kostigen 2020, chap. 1). There is talk, too, of "hurricane slayer" technologies that could bring cooler water to the ocean's surface and potentially prevent or weaken that destructive heat-powered storm phenomenon (Fleming 2021).

If that is the *what* of geoengineering, we still need to address the *why* of it, the reasons anyone might choose to turn to these technologies. Read through enough of the literature and you will find the standard case given for turning to geoengineering to run like this. In emitting greenhouse gases and altering the chemical composition of the planet's atmosphere, humanity has been conducting a grand and dangerous experiment on the earth's environment. And it is an experiment we do not seem able to stop, at least not quickly. Our emissions are the unintended result of committing to energy and agricultural systems that are deeply integral to sustaining a particular mode and quality of life. There is, therefore, a nigh-insurmountable political and economic inertia that remains to be overcome in order to gain the necessary speed in abating emissions. To commit unwaveringly to approaches so mired is to leave a lot of people's lives and well-being at risk, particularly those most vulnerable to the effects of climate change. Furthermore, this grand planetary experiment is also far advanced, a cold fact that must be acknowledged by anyone opposing geoengineering because they are dogmatically harbouring taboos about humans manipulating the environment on a large scale. Whether we like it or not, we are already — and have long been — interfering with the earth's climate system. So why not take a more active role in this experiment and, at the very least, buy some time for those mainstream mitigation approaches (like the ones described in the chapter on neoliberalism) to take hold and for those in the most vulnerable locations to adapt? Why not begin the necessary research now so that we can have a Plan B to deploy once it has become clear to enough people that Plan A is not working?

THE IDEOLOGICAL ASSUMPTIONS OF THE GEOENGINEERS

In this book's other chapters, relating a climate response to a political worldview is relatively straightforward. The responses have advocates — political thinkers or members of political parties — with clear ideological beliefs or, as seen in the last chapter, they appeal primarily to a segment of the political spectrum or a party's voter base.

But geoengineering is unique. It is a solution that appears, or is at least presented as, apolitical and pragmatically realist, and whose proponents do not often openly or strongly assert views characteristic of well-known and well-established ideologies. And so we have to try something a bit unconventional here and identify assumptions and common attitudes evinced when advocates present arguments to validate and justify a geoengineering response and defend it against critics behind arguments used to validate and justify the geoengineering response, and then attempt to situate them politically. To do so, this section draws from three pro-geoengineering books: David Keith's (2013) *A Case for Climate Engineering*, Oliver Morton's (2015) *The Planet Remade: How Geoengineering Could Change the World*, and Kostigen's (2020) *Hacking Planet Earth: How Geoengineering Can Help Us Reimagine the Future*.

A first and most obvious feature is an extremely strong anthropocentric (i.e., human centred) view of the degree of permissible direct human intervention in natural systems. Geoengineering proponents sometimes note that humanity has a long history of intentionally shaping its natural environment; engineering the climate itself is just one more instance of our taming nature. O. Morton (2015, 24–25) sees it as an extension of the *human empire* — Francis Bacon's term for the fusion of scientific power and knowledge applied to the world — which is now itself a force of nature. From this perspective, opposition to geoengineering appears irrational and morally irresponsible, born from ignorance about the historical nature of our relationship with the environment and needlessly circumspect about the growth of human world-manipulating capabilities. Indeed, pro-geoengineering authors note how, if allowed to advance sufficiently, geoengineering technologies might allow us to customize our planet's climate for our collective benefit and even deepen our relationship with the environment (Keith 2013, 173–74; O. Morton 2015, 171; Kostigen 2020, chap. 1).

A second feature to account for is the favouring of what are sometimes called techno-fixes over social change as key to the climate response. This preference is for two primary reasons. The first, as mentioned above, is a sense that the plodding politics of the climate crisis simply makes direct climate intervention simpler and faster. Geoengineering provides a way of moving straight to solutions, circumventing difficult intellectual debates, power plays between conflicting interests, and unhelpful politically driven assessments of the climate crisis. The green left, say geoengineering advocates, has been particularly prone to the latter (Keith 2013, chap. 5; O. Morton 2015, 142). By conflating capitalism and environmental degradation, leftists unnecessarily mount anticapitalist or anti-industrial hang-ups onto the climate challenge when nothing of the sort is needed.

The other reason for dismissing social change is a tendency to see existing social relations as largely unproblematic, and even close to ideal. Unlike in the frameworks highlighted in the system-transforming section of this book, there is no need here to address economic inequalities or deprivation, redress the unjust historical legacies of colonialism, or transcend a wasteful and polluting capitalist economic model. Keith (2013, chap. 5) articulates this clearly: "To sustain claims that an effective response to climate change requires a fundamental reengineering of market capitalism is to deny the fact that liberal market economies have done a far better job of environmental regulation than their competitors." At most, it might be necessary to curtail some of the power and political spending of the major obstructionist corporations, who are here seen to be the true barrier to the implementation of solutions like carbon pricing and regulations. If that can be achieved, the Global North can return to its world-leading status in environmental matters.

A third and related feature is the special status given to the trailblazing future-bringer — the inventor, the innovator, the entrepreneur — elevating them to a principal agent in the climate response; indeed, books by geoengineering proponents tend to be replete with interviews with and anecdotes about these types of figures. Because the popular debate over the right climate response is overwhelmingly dominated by people convinced the solution is to be found in the fraught politics of winning some form of societal change, any advancement on geoengineering cannot be trusted to a broad democratic movement, and is instead reliant on the much smaller group of visionary outsiders who can see the problem and its solutions in terms the rest cannot (or are unwilling to) and who are uncowed by taboos about advancing human sciences and control into still uncharted frontiers. As Kostigen (2020, introduction) said of geoengineering, "This movement won't work at the grassroots level. It is time to turn our collective attention toward supporting industry and encouraging the business community, scientists, and technologists — innovators! — to step up and do what they do best: invent, pioneer, disrupt the same old ways of doing things." There is something unique that people of this sort do for humanity. O. Morton (2015, 27), for example, describes CDR researchers as "the sort of men who make knowledge

— both theirs and, once you learn from them, yours — feel like power. Men of human empire. They are also the sort of men who can attract the interest and admiration of wealthier and more powerful men." It is this minority of men of the human empire (alongside those wealthier and more powerful men) who see the straightest pathway to the future and who will ford us into it — for all our sakes. They are "pioneers ... working on geoengineering projects behind closed doors, deep underground, far out in deserts, in remote jungles, and even in outer space. Brought into league, they can engineer a more benign environment for us all" (Kostigen 2020, 38).

These are not apolitical inclinations. Take the assumption that the climate's complex interacting systems can and should be subject to human manipulation, that these interventions are not just morally permissible but also morally imperative, and also carry a high degree of likely effectiveness (or at least a low likelihood of unforeseeable irreversible damages). The climate crisis is a clear instance of human imposition on the environment gone too far, a new political condition; it is thus for many people a blaring clarion call to reduce our impacts on the intricately interconnected earth system and embrace a politics that leads us to live more humbly with nature. But for geoengineers, this new condition is instead a call for the obverse, to push even further, only this time with confident intent. The need for environmental protection and regulation — and redefinition of our relationship with the planet — decreases insofar as potential damage is mitigated by technology.

Recall that climate change sounds an alarm that something in our society is not working. Skirting past questions of social change, as geoengineers do, muffles that alarm and deprioritizes potential political debates about what aspects of society are not working, and how we can democratically alter our institutions so that they create a more just and sustainable world even as they address climate change.

We see strong hints of having imbibed assumptions in line with the contemporary status quo when geoengineers charge those on the left with dogmatically conflating capitalism and environmental degradation and when they hold out market capitalism (just without the more egregiously obstructionist corporations) as a model of sustainability. There is even a potential positive role for inequality. In a standout passage, O. Morton (2015, 353) contemplates the tendency of billionaires to use their wealth in world-changing ways. It leads him to celebrate the philanthropic vaccination programs of Bill and Melinda Gates as having saved as many lives as were taken by Mao Zedong, Joseph Stalin, and Adolf Hitler, and to note how Elon Musk is changing the course of history by making space travel cheaper and more routine. Morton's political assumptions here are made clearer if contrasted with more egalitarian political beliefs. For instance, social democrats question the system that permitted and encouraged the extremes in wealth inequality that gave rise to these and other members of the multibillionaire class, and argue this is precisely what deprives democratic society of the economic resources needed to rapidly establish a more just postcarbon world. Assumptions like the above are why critics (e.g., Hamilton 2013, 175; Malm 2021c, 147) detect in geoengineering advocates a kind of devotion to capitalist society and its elites, which turns into a conviction that the physical planet is malleable but the social world is not.

Finally, to prioritize the changing of the world through innovations in climate-intervention technologies is to draw the climate struggle away from the world of movements, political parties, and governments — the world of democracy — and towards the home turf of "the business community, scientists, and technologists — innovators!" It recasts the contest over the climate response in their terms, in this case being able to most convincingly promise raw, near-immediate, breakthrough disruptive techno-effectiveness. To the degree those promises are convincing, they render other ideas about social change increasingly moot. Citing catalytic converters that cut automobile air pollution, new refrigerants that would not deplete ozone, and new insecticides to replace DDT, Keith (2013, 152) writes, "Should we prefer social to technical fixes? One can argue either side, but whichever you prefer it's hard to avoid the conclusion that most of the big environmental wins of the last half century have been techno-fixes." In 2021, in order to support a four-year CDR competition, billionaire Elon Musk and the Musk Foundation offered $100 million, allegedly "the largest incentive prize in history," to the charity XPRIZE — which describes itself on its website as "a proven platform for impact that leverages the power of competition to catalyze innovation and accelerate a more hopeful future by incentivizing radical breakthroughs for the benefit of humanity." Perhaps the winner's technology will one day rank among Keith's list of techno-fixes.

SITUATING THE POLITICS OF GEOENGINEERING

By identifying the assumptions built into the case made for geoengineering, we can better ascertain in which political framework it might fit. Probably the strongest candidate is the neoliberal one. The views held by geoengineering proponents are easily accommodated by the reigning political and economic order, and geoengineering can enhance the legitimacy of the main neoliberal climate response. A need to prioritize economic growth and profit means that neoliberal policymakers are unlikely to deploy their main policy tools of carbon pricing and regulations strongly enough to curb emissions in line with what is required to prevent severe climate change. Geoengineering holds the promise of complementing those policies, buying more time for them to drive fossil fuels from the market and reversing some of the climate change that those policies will work too slowly to prevent.

A second candidate is the political right. As the effects of climate change grow even more unignorable and destructive, an increasingly untenable denialism may loosen its grip on the ideological right, and the geoengineering framework is a strong candidate to house the right-wing climate response.

As noted, geoengineering offers to trip up the impetus for the kind of social change favoured by the left. Like denial, geoengineering provides protection to the political and economic status quo and patterns of privilege that conservative individuals value. But it also gives permission to embrace climate science by preventing it from necessarily suggesting social change. In a piece entitled "It's Time for Conservatives to Own the Climate-Change Issue" in the conservative outlet *National Review*, Republican House Representative Dan Crenshaw (2020) argues that the political right could take the lead on climate change by embracing the free market to monetize carbon captured through geoengineering. He says it might even help fight back against what he believes to be the "alarmism" characterizing the climate left and its "ever-more-extreme 'solutions,' " such as the social democratic Green New Deal. (However, the comments section below the piece features readers rejecting the notion that climate change is happening altogether, a sign of the continuing difficulty segments of the right face in accepting the crisis and its severity.)

An embrace of geoengineering might allow those on the political right to turn the tables on the left and, for a change, depict progressives and radicals as the ones who are antiscience whenever they reject direct climate-intervention technologies. Geoengineering elevates a particular kind of science, sometimes called *production science*, that is highly compatible with a right-wing worldview. Science and technology played a crucial role in the sweeping changes that advanced industrial capitalism brought to the world. They granted new powers that gave economic producers novel ways to exploit and manipulate resources, control workers, and even influence consumers. In the Global North, in particular, they drove improvements in quality of life, an expansion of career and consumption choices, and a growth in the military arsenal to defend all of it against tyrannical rival world powers. But the same technological forces that gave us the modern world also had severe unintended consequences, not least among them environmental. In the decades after World War II, new scientific fields came to prominence that were critical of the impacts that industrial capitalism and its technologies were having. New social movements also arose that amplified and made political demands based on the critiques emerging from these *impact sciences*. Together, these forces introduced a new and unsettling countercurrent that complicated the story of modernity, a story that in its simpler versions had promised that capitalism and its allied production science led always and ever to social improvement, and displayed little consideration for the consequences of strong anthropocentric attitudes towards nature.

But if modernity was no longer necessarily unidirectional — if in the long run it could undo itself and its gains as the new impact sciences were intimating — there was no more certainty about what mode of life people should be living under. The preoccupations of modernity suddenly involved not just technical questions about how to expand economic production and consumption but also questions about the ways in which the harms of that production and consumption could be reversed, all of which opened capitalism up to criticisms. Modernity had turned back in on itself — it had become reflexive — and it was revealing the dangers of a society modelled on classical right-wing assumptions that self-interested market actors left to their own devices improve the social good.

Climate change and climate science form the ultimate example of this, and a type of antireflexivity took root on the political right as climate change denialism (McCright 2016; McCright and Dunlap 2010), creating a motivation to reject scientific findings that upend that simple story of modernity. But geoengineering, with its capital-intensive technologies and large role for maverick entrepreneur-innovators bringing their ideas to the market, presents a vehicle for restoring and re-elevating production science (Hamilton 2013, 134), and sapping the urgency out of impact science and the reflexive, system-changing implications it might hold.

Thus, there is reason to think that geoengineering is compatible with a right-wing worldview, particularly if elite conservatives and libertarians get behind it in a committed way. But the political right of the early 2020s is conflicted. There are figures in support, like Crenshaw. There are, too, some right-wing think tanks with a history of asserting climate change denier views now paradoxically showing interest in geoengineering research designed to address the very crisis they would not accept (Hamilton 2013, 90–93, 98–99, 209). But at the same time, people like Tucker Carlson are framing geoengineering as a means for liberal elites such as Bill Gates to impose control over the environment itself as part of some ill-defined ideological project (Halon 2021). These are perhaps signs of a still raging battle within today's right between traditional, small-government free marketers and a more sinister current of conspiracy-fuelled antileftist and authoritarian ethnosupremacism mainstreamed in the mid-2010s.

Recall how ideological frameworks define the structure of a climate response — determining what it includes and how far it can go. Geoengineering enlarges both the neoliberal and the right-wing frameworks, extending the options for ideologically consistent climate responses into the realm of climate manipulation while simultaneously keeping walled off those responses that would see the social realm transformed in ideologically inconsistent ways.

THROUGH A CLIMATE JUSTICE LENS

Critics see in geoengineering a dangerous hubris and paternalism, a potentially cynical ploy to preserve existing political and economic power concentrations, and a set of technologies that are limited by capitalist logic. The geoengineers, meanwhile, often see in their critics a timidity or irrationality that might prevent the saving of lives.

Let's begin with charges that geoengineers are hubristic. Science and industry grant their users a sense of power and control. When technologies are introduced and adopted, it is not with the intent of causing side effects, but rather under the assumption that side effects will not occur, will be inconsequential, or can be easily stopped or counteracted. Any such assumptions are dangerous when it comes to technologies capable of altering the complex, interacting environmental conditions required to protect means of subsistence, maintain human rights and security, and preserve ecosystems. The side effects of interfering with planetary energy distribution, precipitation patterns, or nutrient flows are potentially catastrophic in space and time. This is why it is not entirely convincing when proponents make a case for geoengineering by pointing out that humans have always intervened in their environments. The scope and potency of proposed interventions makes geoengineering nonanalogous. They are a change in the kind, not merely degree, of intervention, "the crossing of a new threshold on the spectrum of environmental recklessness" (Gardiner 2011, 394). As one critic put it, "When uncertainty about the consequences of a course of action is extensive and deep, when the costs of assuming more than we are entitled to given this uncertainty are catastrophic, and when these costs will fall on people separate to those making decisions under these assumptions, we are morally required not to make political decisions using these assumptions" (McKinnon 2019, 10).

But perhaps, an advocate might argue, the side effects of geoengineering could be addressed with further geoengineering. In an interview, Govindasamy Bala, a lead author of the sixth assessment report by the Intergovernmental Panel on Climate Change, noted how changes in precipitation patterns caused by stratospheric aerosol injection might be addressed by cirrus cloud thinning. "The science is there," he remarked (Spring 2021). A first question is, *can* the side effect be neutralized and can it be done without triggering still another side effect? In other words, *is* the science there? Second, who will take up the instances of geoengineering required to offset the side effect? We can imagine scenarios where one powerful and wealthy party undertakes a first deployment of geoengineering to benefit itself, but the side effect harms a party incapable of conducting a second deployment on its own or convincing other parties to do so. The effects of geoengineering will have been externalized onto a weaker party that did not choose to take them on.

A major concern involves proponents' tendency to dismiss the complexity of governing geoengineering research and deployment. One possible side effect often raised is termination shock, a sudden cessation in the application of one of the geoengineering technologies used to prevent some significant portion of sunlight from being absorbed by the earth. For something like stratospheric aerosol injection to work, it would need to be sustained for as long as greenhouse gas concentrations would otherwise commit the world to an intolerable level of warming, meaning regular injections would be required to replenish the artificial stratospheric veil. The danger is that an unforeseen event — terrorist attack, state collapse, environmental disaster, political leaders denying climate change or recognizing no responsibility to other nations — would disrupt one or several points in the globally distributed infrastructure needed to provide the various parts of aerosol delivery (e.g., drones, nozzles, sulphur compounds, monitoring equipment), and a sudden influx of sunlight would lead to a punishing spike in temperatures. Some form of coordination would be required to ensure that, in those worst-case scenarios, others would step forward to maintain the veil. Proponents of SRM have tended to depict this as easy and straightforward, ignoring the serious complexities in ensuring the necessary degree of intergovernmental and intercorporate coordination, trust, and transparency (McKinnon 2019).

In addition to hubris, geoengineering has attracted accusations of paternalism (Hourdequin 2018), the making of decisions on behalf of others who have not offered informed consent, and in a way that undermines their autonomy. When advocates argue that geoengineering should be pursued because it will save the lives of the global poor or vulnerable, the argument is not made with much concern that the global poor or vulnerable actually assent to it; the view seems to be that geoengineering will be good for them, regardless of whether they realize it or want it. What studies exist on geoengineering attitudes of vulnerable populations with a history of oppression show serious hesitance about its use. This is due to concerns that geoengineering will undermine their self-determination, fail to account for climate conditions most important to their livelihoods, leave them with the costs of unintended impacts, and remain under the control of affluent technologically advanced societies (Hourdequin 2018, 278–79).

In some cases, even the way geoengineering is talked about when seeking informed consent starts off on the wrong foot. Scholar of Indigeneity and climate justice Kyle Powys Whyte (2018; 2021) observes there have been attempts to simplify Indigenous Peoples' views on geo-engineering to either validate or delegitimize the technologies, but in reality they are too diverse and complex to fall uniformly onto one side. These views also cannot be separated from conversations about past and ongoing colonial domination that continue to keep Indigenous communities in situations where they cannot meaningfully assent to or dissent from policy decisions proposed from above. As Whyte (2018, 304) states, "A conversation about geoengineering that, say, disallows or is silent on, treaty rights or colonialism, is not a space for Indigenous voices to matter, in my opinion. Or a discussion where Indigenous peoples are asked to trust non-Native people again, this time, is problematic if there are not direct reasons given for why trust is an appropriate attitude. For the conversation must address why distrust occurred in the first place." Even proposing geoengineering as a means of preventing some dystopian coming world ignores how colonial pasts have left many Indigenous communities in what effectively already are dystopias compared to how their ancestors lived (Whyte 2021, 78–79). With some CDR technologies, pipeline networks will be necessary to carry captured carbon dioxide, raising questions of territorial sovereignty and environmental safety similar to those that arise in relation to oil pipelines.

The ethical implications of geoengineering get still more complicated and disconcerting. Situations could emerge where geoengineering responses reduce the total amount of harm compared to a scenario of unabated climate change but radically alter the people who end up being harmed or involve a trade-off in which certain human lives are prioritized at the cost of massive nonhuman life and ecosystem loss. Although the evil being done might be "lesser" in totality, the act of choosing whose lives and well-being to sacrifice would nevertheless remain an unconscionable evil (Gardiner 2011, 392). This sort of dilemma ought to give us pause: Do we actually want the powers geoengineering bestows? Do we have the moral theory (and capacity to abide by it) necessary to make choices about attempting to engineer climatic conditions, particularly where they could involve zero-sum trade-offs of benefits and burdens? In other words, would the capacity of our technology exceed the capacity of our ethics?

Another line of climate justice critique to raise here is that geoengineering is likely to be of greater benefit to the very fossil fuel industry driving the climate crisis than to the efforts to stall and reverse it. We can consider two avenues. First, moral hazards occur when risky behaviour is promoted because some measure has been introduced to reduce the costs of those risks. With respect to geoengineering, one concern is that the technologies will reduce the political urgency of eliminating fossil fuels. As such, geoengineering would not so much extend time for, say, neoliberal solutions to work as it would prolong the life of the fossil fuel industry. Second, geoengineering might be a false solution, a cynical strategy in which the fossil fuel industry makes outward signs of embracing a response that in actuality does little to stop emissions. Industry players might tout nominal investments in CDR technologies like bioenergy with carbon capture and storage or direct air capture in order to present themselves as partners in sustainability, even if the amount of carbon dioxide they remove is an insignificant fraction of the total emitted through production, refining, transport, and eventual burning of fossil fuels. In both cases, geoengineering (or its mere prospect) contributes to further carbon emissions and exposes frontline communities exposed to the health hazards of pollution from fossil fuel extraction and refining.

Finally, the benefits of direct air capture will stay limited as long as they remain under capitalist logic (Malm and Carton 2021). The grand potential of DAC is this: if carbon can be withdrawn from the air, it can then be sequestered, left inert somewhere in the earth. If scaled up, the approach might conceivably even allow us to restore atmospheric carbon concentrations to safe levels. A key problem is that the captured, to-be-sequestered carbon is a difficult commodity to find much of a market for. And so instead of carbon sequestration, DAC companies are embracing carbon utilization. The captured carbon, properly processed, can be utilized, for instance, by soft drink companies to add fizz to their drinks or as a component in synthetic fuels that replicate diesel, gasoline, or jet fuel. The carbon is then just rereleased upon use (one exception is captured carbon dioxide used for concrete). It represents an improvement over adding the emissions of newly dug up hydrocarbons to the atmosphere, but falls far short of direct air capture's potential contributions to restoring a safe climate. And, other than for reasons of climate change, there is little impetus to use air-mined carbon dioxide in products like synthetic fuels; should the high costs involved in obtaining their inputs through direct air capture prevail, it is difficult to imagine these synthetic fuels displacing much traditional fossil fuel in the market

(Roberts 2020b). One sizable market, though, is for enhanced oil recovery, where captured carbon dioxide can be pumped into flagging wells for the purpose of accessing their more recalcitrant drops of oil — hardly an answer to the crisis. Additionally, major DAC pilot projects are still powered by fossil fuels like natural gas. Though there is an assumption that they will eventually be powered by renewables, it raises the issue of whether mitigating fossil fuel emissions is the best use of that still often scarce clean energy.

So there are abundant problems and red flags with geoengineering. But using our justice lens in this late hour also means not dismissing out of hand geoengineers' claims that these technologies will protect lives, particularly those most vulnerable to the effects of climate change. If a country were, for instance, subject to repeated instances of lasting and massively lethal heatwaves, and if the international community showed little sign of lending humanitarian or adaptation support or taking drastic measures on mitigation, would there be clear-cut moral reasons for that country not to deploy SRM technologies? Even if there were, would there be reasons to expect that country would abide by them? Furthermore, CDR technologies could conceivably be used to bring a changed climate closer to conditions that prevailed in the preindustrial period, potentially reducing the potency and reach of climate disruption and probability of triggering disastrous tipping points.

As noted above, geoengineering advocates make arguments that not only fit within the neoliberal and right-wing frameworks but also extend those frameworks outward to permit the inclusion of climate engineering in ideologically consistent ways. They reduce the need for strict environmental regulation, lessen the appeal of progressive social change, and even find a positive role for economic inequality. For parts of the right that have felt the temptations of climate change denial, it even permits a reconciliation with science.

Could geoengineering also allow a framework extension for progressives and leftists? Until now, it has been common to find prominent voices in those camps rejecting geoengineering and expressing instead a profound skepticism about the deeper worldview and system-preserving political project of those advocating for it. For example, democratic socialist Bernie Sanders' 2020 US presidential platform rejected geoengineering as a false solution (see also Schneider and Fuhr [2020]). Many pro-geoengineering arguments would raise suspicions that the endeavour is primarily a way to protect capitalist mastery over society and the natural world. We have yet to examine those more progressive ideological frameworks, but for now let us say that for geoengineering to be acceptable within them, the decisions over the use and development of the technologies should come under control of democratic movements and states rather than technophiles and technocrats. Or, as the respected science fiction author Kim Stanley Robinson (2018) put it, "What we need is science guided by its own scientific methods when doing the science — then guided by a leftist tilt toward justice and sustainability when it's put to use. We need to choose to put science, technology, engineering and medicine to good human and biosphere work, rather than let it be bought to serve profit for the few most wealthy."

Holly Jean Buck (2019) cautions fellow leftists about overly simple framings of geoengineering, one example of which has been to see it through the lens of bad-good binaries casting climate engineering as necessarily opposed to desirable progressive social change: "When it comes to geoengineering, many environmentalists have adopted a simple refrain: 'We don't need geoengineering, we need x.' This is a familiar formula, where x may be sustainable, ecological agriculture. Or system transformation. Or degrowth. Geoengineering serves as a foil for the beautiful x, the blossoming future we really want" (Buck 2019, introduction). To do this ignores geoengineering's potential to alleviate great suffering in the event of runaway climate change. It also ignores the possibility of integrating climate engineering into progressive social transformation. Any serious and successful geoengineering effort, argues Buck, would require a commitment to renewable energy, itself requiring progressive social change. For these reasons, she urges against seeing geoengineering simply as technology, which forfeits decision-making power about its development and applications to a minority composed of technocratic experts. If civil society is to be involved, it is necessary to see geoengineering in different ways — as a process of infrastructure or social intervention. There are, after all, precedents and analogies allowing for people to think critically about changes in these ways and to imagine how to use them to promote the collective good, as well as how values, social movements, and technologies interact to make them more successful (Buck 2020).

Some have argued that carbon-capture technologies should be socialized as part of a leftist political project (e.g., Malm and Carton 2021;

Parenti 2021). The fossil fuel industry can be nationalized and converted to atmospheric clean-up, its chemical and engineering experts and workers now using their skills for the environment while being saved from a dying sector of the economy. Similarly, the state could nationalize some portion of the automotive industry's productive capacity and use it to produce DAC modules. Those efforts can be supplemented if vast lands appropriated for use by the meat industry, itself a significant contributor to climate change, are reappropriated by the state and used for afforestation.

CONCLUSION

As presented by proponents, arguments for geoengineering are highly compatible with neoliberal and right-wing frameworks. Integrated into those worldviews, geoengineering becomes a strategy for attempting to preserve much of the status quo, even possibly for the fossil fuel industry should climate engineering technologies lend validation to the idea that emissions reductions are not urgent. At the same time, attempts to preserve the status quo with geoengineering potentially risk utterly changing the planet should geoengineering come with potent side effects. Those holding more progressive and radical ideologies are left to debate whether to oppose geoengineering or attempt to harness it for social ends. But what if the need for much geoengineering could be eliminated in the first place by truly and aggressively taking on emissions? It is not something the system-preserving frameworks seem capable of doing, and it's worth recalling why before we leave them behind and move on to the system-changing frameworks.

The neoliberal climate response has to contend with its ideological preference for an economy characterized by well-functioning markets, which are believed to distribute society's scarce resources efficiently and optimally and thereby promote the social good. Neoliberals would never go so far as to start planning large sectors of the economy — that would reach beyond where the ideological framework ends. And neither could they rely upon highly deregulated markets, as they normally would, to address the climate crisis. The neoliberal response centres on seeing climate change as a market failure in which the true costs of using fossil fuels are not included in their price. That negative externality can be internalized, primarily through government-mandated carbon pricing and regulations, to make the market function more effectively. But policymakers have shown themselves extremely hesitant to raise carbon prices high enough to bring emissions down to levels consistent with the Paris Agreement goals. Where a choice has to be made between ambitious climate action and protecting economic growth consistent with neoliberal institutional arrangements, the former loses out. Only as much of the world will be saved from climate change as nondisruptive market-friendly policy allows.

In numerous countries, the right-wing response, meanwhile, has been dominated by climate change denialism. The ideal society of the political right prioritizes individual liberty and the protection of property rights, leading to an embrace of free markets and a state that must not intervene in the economy. This system of political beliefs creates a susceptibility to climate change denial because addressing the crisis at anything like the levels necessary will require policies that undermine the ideology's central tenets. Governments will need to intervene in the economy in ways that will affect the liberty of individuals to buy high-carbon products in the market. For decades, a denial machine funded by the fossil fuel industry and politically conservative organizations has spewed ideologically comforting and superficially plausible sounding but scientifically worthless denier arguments and arranged for their propagation through ideologically aligned politicians and news media, all to prevent the social and economic changes that can preserve a livable climate. The operators of the denial machine have engaged in one of the most morally execrable endeavours in human history.

For all their disagreements, these frameworks share some important features. Green political theorist Robyn Eckersley (2004, 108) once wrote of several "liberal dogmas" that have long stymied an ecological shift within the dominant order and that apply to what we have seen, liberal or otherwise:

> a muscular individualism and an understanding of the self-interested rational actor as natural and eternal; a dualistic conception of humanity and nature that denies human dependency on the biological world and gives rise to the notion of human exceptionalism from, and instrumentalism and chauvinism toward, the natural world; the sanctity of private property rights; the notion that freedom can only be acquired through material

plenitude; and overconfidence in the rational mastery of nature through further scientific and technological progress.

Perhaps their most important shared feature, however, is a preference for maintaining a sharp separation between the economic and the political. The matter of who gets what and how much of it is mostly a matter left for capitalist markets to figure out. There is little interest here in creating space for people to make collective, democratic decisions about the economy. The system-changing frameworks that we are about to look at, for all their differences, have in common a radically different ideological belief that an ideal society is one that allows for democratic shaping of the economy with regards to how it should function, how large it should be, what investments might be made with the enormous surplus it generates, and in whose interests. By enlarging the role for economic democracy, these frameworks offer quite different climate responses that also open opportunities to address major questions of justice. They contain possibilities for much different worlds being ushered in than those offered by the system-preserving frameworks.

PART 3

....

THE SYSTEM-CHANGING FRAMEWORKS

7

SOCIAL DEMOCRACY AND A GREEN NEW DEAL

A GREEN NEW DEAL GOES TO WASHINGTON

DESPITE THE OVERCAST DAY, a bright and piercing vision of a future world arrived on the American political scene on February 7, 2019, as one of the Democratic Party's rising stars, Alexandria Ocasio-Cortez, alongside veteran Democrat climate hawk Ed Markey, put forward a new policy resolution: the Green New Deal.

The resolution marked some breathtakingly ambitious goals for the US political context. Chief among them was that of supplying 100 percent of domestic energy needs through clean, renewable, and zero-emission sources in ten years. But there was so much more for Americans: job guarantees (with living wages and paid vacations), education for all "with a focus on frontline and vulnerable communities," the obtaining of free, prior, and informed consent from Indigenous communities with respect to decisions affecting those communities and their traditional territories, and the provision of "high-quality health care; affordable, safe, and adequate housing; economic security; and clean water, clean air, healthy and affordable food, and access to nature."

Though it took the form of a nonbinding resolution (and was voted down in the Republican-controlled House of Representatives), the effort by Ocasio-Cortez and Markey did what it needed to: massively changed the national conversation around climate policy. As many media outlets and analysts observed, it succeeded in hurling open the Overton window, the range of policy being seriously considered in the mainstream, with regards to responding to climate change. The arrival of the Green

New Deal created space to discuss a far more expansive and ambitious response to the climate crisis at a time when mainstream discourse of climate policy had largely been limited to which choice of carbon-pricing measures to institute. And even that conversation had come to a screeching halt.

The presidency of Donald Trump was a major setback for climate policy in the United States but also a powerfully galvanizing moment for the climate movement. In some of the last years left to act to achieve the Paris Agreement goals, a man who denied the crisis altogether was elected to the highest office of power. (His bragging about the state of America's clean air and water when pressed on the matter of climate change suggested he did not even understand what climate change was [Bump 2019].) It was a blaring signal that something bold would be needed to make up for the time lost and to shake up the political order that had produced Trumpism, something that could resist being clawed back by future obstructionist presidents or a Republican-controlled Congress. In this context, the Green New Deal was understandably greeted with much enthusiasm by those concerned with advancing climate justice.

But why should responding to climate change have anything to do with the government securing rights to housing, economic security, health care, and education? To understand the ideas behind the Green New Deal (and related programs around the world), we need to understand it through an ideological lens.

HUMAN DEVELOPMENT: SOCIAL DEMOCRATIC IDEOLOGY

The chapter on denialism highlighted the dual meaning of freedom, sometimes distinguished using the terms *freedom from* and *freedom to*. The first — the narrower view — is the sense of freedom of the political right, which leads it to favour a small-government, free-market capitalist society. This is because what reduces freedom is coercion — that is, credible threats of force. The state's role is intended to be minimal, protecting people from being coerced and cheated by others while also refraining from creating laws unjustifiably forbidding people's actions. In an economically right-wing society, people may be *allowed* to do many things, in that these things are legal and people can pursue them free from fear of punishment by the state or interference by others. The problem is that they may not be able to actually do them for reasons of material deprivation, absence of opportunity, and so forth (Meyer with Hinchman 2007, 102).

And so the second view of freedom includes concerns about opportunity and ability. Freedom, according to this more expansive sense, is reduced not only when people are coerced by threats of force but also when they cannot access goods or services required to actually execute their plans of life. They must have *rights* to these essential goods and services themselves. Social and economic rights must be added to political and civil ones. The social democratic bundle of fundamental rights tends to include rights to sustenance, housing, health care (including dental, pharmaceutical, rehabilitative, eye, and mental health care), education (primary through postsecondary, trade school, or graduate school), child care, employment and a livable wage, and leisure. And, as society evolves, the list need not remain static. In recent years, social democrats have come to advocate for rights to quality internet service and even a universal basic income (e.g., Bregman 2017; Standing 2017), a regular payment to every person regardless of employment status that is enough to lead a decent life.

These rights are intended to guarantee not only freedom but also, in their universality, equality. They define a minimum standard of life to which all members of a society have equal claim and that is set high enough that everyone, even those at the bottom, can reasonably be said to have enough to live a free, decent, and dignified human life. Prominent examples include the United Nations Universal Declaration on Human Rights and US president Franklin Delano Roosevelt's proposed Second Bill of Rights in 1944, made under the conviction that "necessitous men are not free men."

As this book turns to the system-changing frameworks, it will be useful to better understand the notion of *human flourishing* as a standard of well-being and freedom. This standard is based on the view that an essential feature of human nature is that each person has an array of things they want to do, experience, learn about, achieve, discover, complete, master, experiment with, and so forth — all within the limited time they have on earth. (Or as poet Mary Oliver wrote, "Doesn't everything die at last, and too soon? / Tell me, what is it you plan to do / with your one wild and precious life?") A good society, it follows, is one that allows its people to maximize the combination of realistic opportunities they have concerning what they value doing and achieving. This is known as the capabilities approach to justice, proposed by philosopher and development economist Amartya Sen and further developed by philosopher and liberal social democrat Martha C. Nussbaum (2013, chap. 2).

How can this vision of the world be made real? The experience of the neoliberal period is all the confirmation the social democrat needs that the answer is certainly not to leave capitalism to its own devices. Capitalism does not, after all, provide services or goods when it is unprofitable to do so; those who cannot afford to pay the market price for something must simply go without. What real *right* to education (including postsecondary and graduate schooling) or health care (including prescription medication, dental, and mental health), for instance, do people have if they cannot afford it? And there exist practices under capitalism that are predatory and destructive, leading to environmental degradation (through pollution or overuse of resources), worker exploitation and poverty, and even system-wide financial crises.

And then there is the inequality. The neoliberal economy works overwhelmingly and staggeringly in the favour of those at the very top. Since the start of the neoliberal period, the gains of economic growth have been captured by the wealthiest at scales that most of us cannot fathom. Consider wealth, the value people hold in their bank accounts, property, stocks, and so forth. According to Credit Suisse (2020), the richest 1 percent of humanity, the fifty-two million adults holding a million dollars or more, control 43.4 percent of the world's total wealth. The bottom 54 percent of humanity hold 1.4 percent. In Canada, the top 1 percent control a quarter of the country's wealth while just the top 0.5 percent control a fifth of it (Office of the Parliamentary Budget Officer 2020). In the United States, the top 1 percent control 30 percent of household wealth while the bottom half control not quite 2 percent. If the more equitable postwar pattern of income distribution between the top earners and the rest had prevailed, then $47 *trillion* that went to just the richest 10 percent between 1975 and 2018 would have gone instead to the bottom 90 percent, according to one study (Price and Edwards 2020).

For those even further to the political left, these are reasons to reject capitalism altogether. But social democrats note that capitalism also has important upsides. The economic dynamism inherent in it provides an ever-expanding horizon of possibilities through new products and services being invented and brought to reality. And like no other economic system, it generates enormous economic surplus that can, properly harnessed, raise a society's quality of life.

And so social democrats advocate for a large role for the state in managing economic affairs. Careful state regulation can cut off capitalism's more predatory and crisis-triggering strategies for pursuing profit. Minimum- or livable-wage laws, for instance, prevent firms from paying poverty wages as a means of increasing profitability. Regulation and state oversight of the finance sector can, similarly, close off practices that might drive short-term profit but risk system-wide instability and potential crisis of the kind seen in 2008–09. Through careful and forward-looking industrial policy, governments can steer the economy away from reliance on fading industries towards fuller employment in more dynamic sectors with longer-term futures.

And, crucially, the state can also promote economic justice through policies aimed at reducing inequalities that occur within capitalism. This starts with the social democratic rejection of the right-wing ideological belief that people are entitled to all of what they inherit or negotiate for themselves in the market (minus taxes to fund very minimal state functions). Progressive state taxation policy can tap into the economic surplus that laissez-faire capitalism concentrates at the top among corporations and the wealthy, and redistribute it to promote the social good. Among sectors that social democrats tend to prioritize are education and health care. It is, after all, difficult to pursue one's life goals while lacking necessary base skills and knowledge or while in ill health. And of course, what can be made part of the public sector can extend beyond health care and education to include housing, transit, social security, and more.

WHAT MAKES A GREEN NEW DEAL
A GREEN NEW DEAL?

The term *Green New Deal* has been around for some time, and has referred to a number of different projects (Schlosser, 2020; Mastini, Kallis, and Hickel, 2021, 2–3). While advocates for a Green New Deal must be allowed some room for disagreement, the term becomes meaningless if it simply describes any (seemingly) big policy response to climate change within capitalism. To better appreciate today's Green New Deal, it is useful to recall what the original New Deal attempted to do. Like Roosevelt's 1933–39 program, a proper Green New Deal ought to have three characteristics, showing ambition equal to the crisis, system transformation, and a connection to justice movements.

Let's begin with how a Green New Deal must be one in which its *ambition is equal to the crisis*. This means first and foremost the pursuit of a program to limit warming to well below 2°C — and preferably below 1.5°C — relative to preindustrial levels. That will entail very drastic reductions in fossil fuel emissions starting immediately. Many climate plans today aim for net-zero emissions by 2050 without ambitious targets in intervening years, the danger being that governments will neglect to do much in this crucial period. A Green New Deal intentionally puts the onus of action on us today. No excuses.

The original New Deal program marshalled massive public investment and social programs to respond to the crisis of its age, the Great Depression. That meant using the resources of the government to put people back to work building the country in the public interest. Roosevelt's administration identified a variety of projects that would employ labourers and contribute to the public good, and to roll them out it created public agencies known by an alphabet soup of abbreviations. Agencies such as the Tennessee Valley Authority (TVA), the Civil Works Administration (CWA), the Public Works Administration (PWA), the Works Progress Administration (WPA), and the Civilian Conservation Corps (CCC) put Americans to work electrifying rural areas, creating public art and music, restoring landscapes, and building or repairing schools, city halls, docks, dams, courthouses, public roads, and parks. Not all of the financing came from government. In some instances, government programs were intended to stimulate private-sector spending and investment. In addition to drawing on responses to the Great Depression, Green New Deal–type programs also draw on wartime national mobilization metaphors and comparisons (Delina and Diesendorf 2013; Delina 2016; Roberts 2016; McKibben 2016; Silk 2019; S. Klein 2020) to reemphasize the kind of response — large-scale, government-coordinated national economic mobilizations — that will be required.

Turning to these historical examples shows that past existential crises have been met successfully with ambitious, government-led, and publicly financed programs. Today's crises can be too. The comparisons to the past are not perfect, however; there is acknowledgement that these chapters of history failed to address historical oppressions, chiefly sexism and racism, and that any Green New Deal will have to avoid similar failures.

But there is more to a Green New Deal than large ambitions. Something has long prevented our societies from responding to climate change at that scale. The second feature of any prospective Green New Deal is that it is *system transformative*. This feature is about recognizing that the climate crisis is rooted in a systemic failure that is far greater than a market externality, than something that can be resolved easily while changing society as minimally as possible. That starts with identifying the dominant system relentlessly driving the climate crisis and preventing it from being attended to at anything like the scale and speed necessary: neoliberalism.

It was a matter of cruel timing that the ascendance of neoliberalism happened to coincide with the years in which decisive climate action had to be taken. Any social order based on accepting the terms set by neoliberal ideology would always be dangerously constrained in what policy tools it could bring to bear on emissions in anything like the time needed; it is like getting wrapped in a straitjacket as a fire spreads. Starved of public funds thanks to lowered corporate and income taxes, governments were stripped of an array of measures that might have long ago galvanized the response, such as investing in large-scale renewable energy infrastructure; providing funds for research, investment, and deployment of green technologies; expanding affordable public transit; or assisting workers to transition out of high-carbon jobs. Neoliberal rejection of state economic planning prevented rapid and planned phaseouts of fossil fuel and other high-carbon sectors, while taboos on environmental regulation and, until recently, pollution pricing kept the dirtiest industries economically viable and their products artificially cheap. These same industries then set their vast wealth to work corrupting the political conversation on climate change, whether by financing climate denier politicians or climate denier think tanks, to prolong their rapacity as much as possible. The same ideology behind the breakdown of the climate is also behind the fraying of the social fabric. For so many people, the world that neoliberalism has ushered in is one of meaningless, precarious, and underpaid (over)work; stagnant wages; indebtedness; eroded public services or increasingly inaccessible and privatized versions of them; and stalled, diminished life prospects.

Any Green New Deal thus entails an ideological fight, one that fully rejects the notion that it is natural, immutable, or desirable that the state should play an increasingly small role in promoting the public good. Instead, a Green New Deal openly embraces the idea that a truly democratic society is under no obligation to quietly accept an economic system that fails to serve its people even as it fails to prevent the destruction of the climate; if it could be shaped to serve the interests of a powerful few, it can be reshaped to serve the rest of us. Real democracies can reconceive what a good society for everyone means — what rights and opportunities to flourish it will guarantee to all — and use the economic power of their governments to achieve it. Finding ways to assert this belief through a policy framework has been perhaps the great innovation of the Green New Deal.

This revivified role for the state in not only taking actions to deal with the crisis of its age but at the same time advancing economic justice is another major reason for evoking the original New Deal. Roosevelt's program is remembered today not just for simply attempting to address the Great Depression but also for having ushered in, if incompletely, a fairer American society as the new normal, with government using its power to support the larger public good. And so for a project to be worthy of the mantle of "New Deal," it has to deliberately seek to bring about a better society on the other side of the crisis, one that asserts values of justice, guaranteed by a democratic state, in opposition to an unjust prevailing order.

It is therefore no accident that Green New Deal–type programs contain policy goals that extend beyond emissions reductions into social and economic domains of education, housing, health care, and the like. These demands stand as deliberate and defiant repudiations of neoliberal principles. They are intentionally unreasonable within the context of neoliberalism but perfectly reasonable from a human rights or human flourishing standpoint. They draw attention to the failings of neoliberalism. That these demands have to be made at all delegitimizes the system failing to realize them.

But neoliberalism is just the latest in a long line of oppressive historical episodes that have taken a generational toll. There are communities to this day living with the legacies of genocide and settler colonialism, slavery, and racism, which the original New Deal failed to address. Today's economic mobilization programs patterned on the New Deal thus become opportunities to also redress long-standing historical injustices (Coleman 2019). If a society is to deploy the resources of the state to promote the social good, it ought to ensure that the frontline and underserved communities who have been historically excluded from economic development programs are prioritized. Otherwise, what is to stop history from repeating itself?

All in all, the results are boldly positive visions of a kinder and much more decent society that works for everyone and that everyone is invited to help build. (See, for instance, the instantly viral *A Message from the Future* video produced by The Intercept and narrated by Alexandria Ocasio-Cortez.) If the mainstream liberal response to climate change is about aligning climate action with profit and economic growth, a Green New Deal is about aligning that action with social justice (Gunn-Wright 2020), using it to repair the "intersecting brokennesses" (Klein 2020) of the environment and society that have arisen through forty years of neoliberal austerity. Climate and justice are made inseparable.

Third, and related to the previous point, the content of any proper Green New Deal *emerges from justice movements* — climate, racial, migrant, food, housing, and otherwise. Unlike the mainstream liberal framework, the change agents of a Green New Deal are not government technocrats who are largely free to interpret what the climate movement is demanding or to pare down those demands as needed to fit within ideological bounds. The policy contained within a true Green New Deal should be bottom-up, created through historical moments of disparate advocacy, struggle, and resistance. It links together a diverse coalition of people who have been fighting against the failures of a neoliberal order. A true Green New Deal would show endorsement by major justice organizations who are also the ones who can recognize if governments are pursuing it in good faith.

MAKING IT REAL

This section focuses on three priorities that will need to be met to make real (and protect) the green energy and good jobs aspects of the Green New Deal in the world. Bear in mind that a Green New Deal would also seek to constrain and phase out fossil fuels through regulations like those seen in Chapter 4 as well as much more ambitious ones that would, for example, put an end to drilling leases for industry.

Paying for It: Reclaiming the Economic Surplus, Rejecting Economic Orthodoxy

By one estimate, a global Green New Deal would cost an average of US$4.5 trillion per year between 2024 and 2050, or an average of 2.5 percent of global GDP between 2021 and 2050, mobilized between private and public actors (Pollin 2020a, 422) — a tall order, but not impossible. Importantly, proponents of a Green New Deal reject the right-wing and neoliberal belief that states must necessarily be strapped for cash, arguing that this condition is a result of deliberate social engineering over the past generation to ensure that economic surpluses are directed to and concentrated among a vanishingly small segment of elites. Or to put it another way, the rich are commanding wealth that is not justifiably theirs. At a certain level of wealth, no one can reasonably be said to *need* more, and the flip side of hoarding it is the deprivation of essential rights for a large number of people. In the context of the climate crisis, arguments for allowing obscene inequality ring hollower still, given how that mass of wealth is not being put towards an existential fight. Still another justification to consider when targeting the wealthy is that they bear an outsized responsibility for global carbon emissions; the richest 10 percent were responsible for 50 percent of global emissions between 1990 and 2015 (Oxfam 2020b).

And so paying for a Green New Deal starts with reclaiming society's economic surplus so it can be redirected towards more socially and environmentally beneficial projects. Part of financing such a program could come from raising marginal tax rates — the amount of taxes paid on each dollar earned above a given amount — on the highest brackets, as Ocasio-Cortez has proposed (Yglesias 2019). And income is only one part of the picture. The rich also hold enormous assets in property, bank accounts, stocks and bonds, vehicles, jewellery, and so on, and thus higher taxes could also be levied on wealth (Saez and Zucman 2019a). In 2019, both Bernie Sanders and Elizabeth Warren included wealth taxes in their Green New Deal platforms. Taxes on large inheritances are still another option.

The neoliberal shift also involved a reduction in corporate taxes both by lowering the effective tax rate and allowing a series of loopholes and workarounds through which corporations could avoid paying their fair share. One particularly lucrative strategy has been to shift their profits to low-tax jurisdictions, or tax havens. Raising national and subnational corporate tax rates is one obvious way to reverse the tendency towards undertaxing corporations. Meanwhile, to address corporate use of tax havens, one promising proposal would force corporations to pay "tax deficits" to governments in their home countries (Saez and Zucman 2019b; Clausing, Saez, and Zucman 2021), the difference between the rate a multinational corporation pays on the profits it registers in a given tax haven and the rate it would pay in its home country.

Funding can also be found from reducing or eliminating spending in other realms inimical to a sustainable and fair society. An obvious one is fossil fuel subsidies (Atkin 2021a) like allowing companies to deduct drilling costs from taxes. And there are subsidies to the fossil fuel industry that extend beyond what goes directly to oil companies. Bernie Sanders's Green New Deal program called for a reduction in military spending, much of American military expenditure being dedicated to controlling the world's fossil fuel resources (Roberts 2018b).

Revenues from carbon pricing are another potential massive source of funding. Green New Deal proponents therefore tend to strongly reject revenue-neutral approaches to carbon taxes and favour higher carbon prices than those found under the more mainstream approaches. Another proposed source of revenue is through financial transaction taxes. Many of those transactions are put towards financial speculation, with dubious benefits for the larger economy and often destabilizing effects.

Making renewable energy generation publicly owned would ensure people's utility payments do not end up in already overfilled private coffers and instead go towards the public good (Bozuwa and Alperovitz 2019). One common critique of private energy utilities is that their profit motive leads to harmful or burdensome and even dangerous cost-cutting measures (e.g., see Aronoff et al. 2019, chap. 1). A well-run green public utility could address that problem. And on the topic of public ownership, the fossil fuel industry could also be nationalized (Pollin 2022) with the profits it generates (as it is phased out) directed to building postcarbon infrastructure.

A Green New Deal would also promote reduction of spending on other state programs. The growth in good, well-paying jobs would reduce the need for social welfare programs. As people are employed in an expanding green economy, fewer will have to rely on unemployment insurance, and other supports for the unemployed, underemployed, and working poor. It would also add government revenue through income taxes on workers.

There has also been increasing attention paid to the role that public banking can play in providing financing for a green transition (Marois and Güngen 2019; UNCTAD 2019). Private banks prioritize the financing of projects that are most likely to maximize financial returns on investment in the short term. In practice, this means they avoid certain kinds of projects, like those with longer time horizons; those in still emerging sectors; those coming from agents who are smaller, newer, lacking substantial collateral or credit histories, or representing underserved communities; and those that promote the public interest but in ways that are not expressed in monetary terms (i.e., positive externalities). Publicly owned banks can step in to help remedy the underinvestment resulting from overrelying on private, for-profit finance, and they have historically played a crucial role in economic and community development because they were given broader mandates than just rapid profit maximization. Within the framework of a Green New Deal, the mandate of public banks would be to make financing available to promote a swift and just transition away from a high-carbon economy. This could be done, for example, by providing loans at below-market rates on either a for-profit or not-for-profit basis; issuing grants, transfers, or subsidies; taking a direct equity stake in projects; or offering in-kind technical assistance and expertise.

And then there is probably the most hotly debated approach to paying for Green New Deal projects, which argues that much of the above is missing the point. The approach is called modern monetary theory, a heterodox economic approach that argues governments do not need to pay for state funding by raising taxes (Kelton 2020). Where they issue and control money, governments can generate it without, as orthodox economists fear, necessarily triggering inflation. And, just like they have done to finance wars, corporate bailouts, Apollo programs, and tax cuts for the rich, they can explore using it to fund a Green New Deal (Kelton, Bernal, and Carlock 2018; Gunn-Wright and Hockett 2019, 11).

Working for It

A just transition for workers is another vital element of a serious Green New Deal. Any rapid transition will mean that high-carbon sectors of the economy can employ fewer workers, the fossil fuel industry being the most obviously affected. Proponents of a Green New Deal argue that the work required to transition to a greener, postcarbon economy will generate a massive wave of employment. This is due to the simple fact that building a new economy will require a lot of precisely that — *building*. Retrofitting homes and buildings to upgrade energy efficiency and constructing infrastructure to generate and transmit renewable energy — all at the scale and speed required to address the crisis — will be highly labour-intensive. This work cannot be done using labour-saving strategies like automation or machinery. The growth in jobs in green industries will outpace the loss of jobs in the fossil fuel sector, and may also be able to absorb some of the latter's displaced workers. In Canada, jobs in renewables began exceeding those directly provided by the oil sands in 2014 (CBC News 2014a).

But more will be needed. Green New Deal proposals have included a number of potential policy tools to ensure no worker is left behind (Pollin and Callaci 2019; Brecher 2020b). Workplace transition plans would require, for example, energy utilities eliminating coal generation plants to lay out plans for affected workers. Wage guarantee initiatives would ensure that workers transitioning into new careers do not experience losses in income. Where new jobs offer lower pay, government programs would fill in the difference in earnings between current work and what was earned in the fossil fuel industry. Education and job training initiatives would provide workers with the opportunity to acquire or improve skills (upskilling) needed to move into new sectors. At their most ambitious (e.g., as seen with Sanders's plan), these initiatives would cover not only full schooling costs to complete a degree or training program but also living expenses over that period. Pension and benefit support initiatives would aid workers in the industry who are not quite at retirement age but old enough to have difficulties starting a new career, and others whose pensions and benefits are tied to companies that would soon no longer be operating. Green New Deal solutions have included providing early retirement to workers who choose it with pensions guaranteed through public initiatives.

And there are other jobs out there still. Green New Deal supporters point out that there are existing sectors of the economy that promote the social good and are already low carbon, like health care, child care, elder care, the arts, public-interest journalism, and education. If better funded, they could not only provide employment but also deliver improved services — smaller classrooms, fewer delays in medical treatment, more responsive concern for the elderly, investigative journalism holding the powerful to account, and more.

Other jobs will be created through environmental reclamation initiatives that restore ecosystems that have been disrupted in the pursuit of fossil fuels. These could entail cleaning up abandoned oil wells and decommissioning the plants, refineries, pipelines, and storage facilities that will cease to be needed in a postcarbon world. Other work might be found in repurposing existing capital for greener uses — for example, using shuttered or underused assembly plants to manufacture electric vehicles or renewable energy infrastructure.

Green New Deal supporters also emphasize the need to ensure these new jobs are good, unionized ones. One of the blows that unions suffered in the neoliberal period was due to corporations offshoring work to countries in the Global South with cheaper labour and weaker labour and environmental regulations. But because so many of the jobs created through a Green New Deal would absolutely need to be local — building renewable energy infrastructure, retrofitting houses, repurposing local capital, restoring damaged environments — the main strategy of breaking unions through offshoring becomes useless. And the need to publicly finance Green New Deal jobs opens up opportunities for unionization that would not be available through private-sector-only jobs.

Fighting for and Protecting It

Transformative, ambitious, and justice-based climate programs of the Green New Deal type are subject to sustained attack. Predictably, those furthest to the right have been the most deeply opposed to them. Any program like the Green New Deal is an existential threat to any political and economic order that protects the wealthiest echelons of society. Compared to the climate crisis, whose effects the wealthy can weather (or at least believe they can weather and still stay wealthy and powerful), it is the politics behind justice-based system transformation that are the more immediate danger.

Predictably, forces on the ideological right in the United States hated it, and Republican politicians and right-wing media pundits assaulted the Green New Deal on multiple fronts. They depicted the Green New

Deal and its proponents as incompetent, naively utopian, and unserious (Media Matters Staff 2019; Shapiro 2019; Swaminathan 2019); grossly exaggerated the costs of the program (Anderson 2019; Colman 2019; S. Waldman 2019); argued it would lead government to strip away individual liberties (Atkin 2019; Rupar 2019); suggested its adoption would cause the collapse of modern civilization (J. Johnson 2021); and even turned to the tired anticommunist scare tactics (or red baiting) that infused political discourse for much of the twentieth century (Donohue 2019; McConnell 2019a, 2019b).

Voices on the right were not alone. Chapter 4 discussed how mainstream liberals attempt to prevent climate responses from transforming society. That was on display in their reaction to the Green New Deal. Democratic representative and House speaker Nancy Pelosi dismissed it, saying it "will be one of several or maybe many suggestions that we receive. The 'green dream' or whatever they call it, nobody knows what it is, but they're for it right?" When young Sunrise Movement activists asked Dianne Feinstein to support the Green New Deal resolution, the veteran Democratic senator was also dismissive, saying there was no way to pay for it and that it would never pass a vote in Congress (McKibben 2019b).

The program's perceived feasibility is the main point of contention among neoliberals. As noted, a hallmark of the ideology is a kind of economic rationality that starts with accepting conditions of public austerity as necessary. Governments may not turn to substantial spending in order to usher in progressive change; any such change, if it is to happen at all, is to arrive through the workings of the market. In some extreme circumstances like climate change, governments may tinker with the market to ensure it sends more accurate price signals, but any program requiring large public investment is, by definition, infeasible.

Under the headline "The Green New Deal Would Spend the U.S. into Oblivion," one columnist for Bloomberg media applauded the Green New Deal for its boldness on climate change but condemned the inclusion of unrelated economic securities, fearing that the reliance on untested modern monetary theory to ensure those securities would be untenable (Smith 2019). Others speculated that the Green New Deal was more of a strategy for raising political levels of ambition and less a serious or literal vision of a workable society. As one piece in the liberal online outlet Slate put it: "The tactic of the moment — of knowingly endorsing daring ideas that do not work as tactic — strikes me as much less clever than its adherents would have us believe. If we no longer think of actual feasibility as a best practice, it is less likely that our actual policies will somehow average out into bold yet feasible improvements on the status quo and more likely that we'll simply be stuck with failure" (Pesca 2019). Similar fears were expressed by neoliberal voices in Canada (e.g., Coyne 2019) as the Green New Deal informed the election platforms of the country's more left-leaning parties.

GREEN NEW DECADE?

The 2020s began with the potential to become the decade of the Green New Deal. The COVID-19 pandemic upended the global ideological context and, with it, the sense of what is politically feasible. It created a state of exception that social democrats and progressives used to expose the inability of neoliberal norms to respond to the crises of our time. Widespread demands were made for a just recovery, government-led responses that would revive the global economy once the pandemic had been handled and the lockdowns that were required to keep the virus under control were lifted. Governments across the world announced plans to increase public spending substantially, with much of it for responding to the climate crisis.

The major context to watch for a Green New Deal is the United States. There, COVID-19 not only created conditions in which Keynesian economic policies were necessary but also drove regime change. The Trump presidency had brought to power a combination of callousness and science denialism, one that translated into an administration as indifferent to climate change as it was to the pandemic. Trump's mismanagement of the pandemic was widely seen as the major factor in his election loss. At the same time, the growing concern about climate change and the popularity of the Green New Deal pressured Democratic presidential candidates Jay Inslee, Elizabeth Warren, and Bernie Sanders to include unprecedentedly ambitious climate policy in their platforms, with the latter two also running on a call for medicare for all. Climate journalist David Roberts (2020a; 2021a) dubbed the climate content of these policies SIJ: standards, investments, and justice. At their core (and achieving rare unity on the broad American left) was a sense that the government could act in a big way on the climate crisis. It could, for instance, set bold standards specifying what percentage of the energy that utilities provide must come from renewables, and make enormous public investments in the power grid, electric vehicle charging stations, and renewable energy. It could foster the creation of union-friendly jobs, helping workers to transition out of high-carbon sectors. And it could promote environmental justice for communities — often low income, Indigenous, or of colour — disproportionately facing environmental harms.

Though Joe Biden ultimately won the Democratic nomination, he and his team sought to close the rift that had opened up in the Democratic Party and its base between establishment neoliberal centrists and the growing progressive movement, going so far as to form a Biden-Sanders Unity Task Force. The consequence was that Biden's platform in the presidential race against Donald Trump was much more ambitious on climate than he had initially proposed, even if it was missing key components of the Green New Deal such as medicare for all and job and housing guarantees (Berardelli 2020), as well as aggressive restrictions on, or plans to phase out, the fossil fuel industry. Green spending targets, for instance, rose considerably compared to what Biden originally proposed, though they did not reach as high as in Sanders's plan.

Once in power, however, the Biden administration failed to realize much progress on climate. Its agenda was balanced in a strange, precarious space between neoliberalism and social democracy. Though it showed willingness to propose public expenditure and standard setting to fight climate change, it also made concessions to the fossil fuel industry. Biden's government approved more oil and gas drilling leases on public lands in its first year in power than did Trump's. In a bit of outrageous timing, mere days after COP26, the 2021 United Nations climate negotiations, the Biden government held what media reported to be the nation's largest ever auction for oil and gas drilling leases, these ones located in the Gulf of Mexico. The leases were cancelled in early 2022 when environmental group Earthjustice won a lawsuit against the government — hardly events one would expect under an administration serious about climate change.

The deep dysfunctions of American democracy — now deep enough to have consequences for the global climate — were also to blame. The main piece of legislation containing major climate initiatives was the Build Back Better Act, which required passage in a senate where Republicans held half of the one hundred seats despite representing, by some estimates, over forty million fewer people than the Democrats did (McCarthy and Chang 2021; Millhiser 2021). This malapportionment is due to an archaic constitutional rule that provides each state two senators, leading to smaller (and today, more conservative, Republican-voting) states holding outsized power relative to their population. In this context, all fifty Democrats and Independents would need to vote for the Act, with the vice president casting a deciding vote. But conservative-leaning Democratic senator for West Virginia Joe Manchin, the largest senate benefactor of election campaign contributions from fossil fuel companies and holding millions of dollars in personal fossil fuel assets (Milman 2021), threatened to withhold his support unless a number of items were whittled down — or eliminated, as was the case with a clean electricity standard that experts expected would be the single greatest contributor to emissions reductions (Roberts 2021a). Even after those concessions, Manchin finally announced in late 2021 that he would not support the Build Back Better Act should it come to a vote, effectively killing the package, which would have included over half a trillion dollars in climate spending. Reflecting on these events, Varshini Prakash, the executive director of the Sunrise Movement, which had tirelessly organized to bring the Green New Deal onto the political agenda, perhaps summed it up best, calling it an "excruciating year" (Krieg and Nilsen 2022).

THROUGH A CLIMATE JUSTICE LENS

Even if it has yet to be fully implemented by a government, the Green New Deal's arrival onto the political scene was a godsend for those concerned with climate justice. It is the first framework we have examined thus far that arises from a democratic movement and deliberately foregrounds matters of justice. In its concerns to achieve the 1.5°C warming target can be seen important intergenerational justice elements. Applying a climate justice lens involves a search for opportunities to challenge aspects of our society that are not working or are indefensible. Green New Deal projects reverse inequalities and prioritize frontline and underserved communities. It is absolutely essential that they are advanced and, wherever they take root, protected and enhanced.

However, for those who feel that capitalism is inherently an immoral system, Green New Deal–type economic mobilizations can leave something to be desired. Let's look at two concerns that arise through a justice lens: the possibility of "green colonialism" and the problem of consumption and sustainability.

Neocolonialism and Capitalist Exploitation

In 2020, Business & Human Rights Resource Centre released a report noting that major renewable energy companies were failing to protect human rights in their policies and practices. The centre had recorded 197 allegations of human rights abuses in the renewable energy sector since 2010. One major concern is that a Green New Deal, however unintentionally, could be a vehicle for a wave of what is sometimes called green colonialism (Gebriel 2019; Rehman 2019). That could occur in a couple of ways.

First, demand for land for renewable energy projects, like large-scale wind or solar, could threaten people's sovereignty — particularly Indigenous communities' — whether through land grabs or disruption of local ecosystems. Members of Saami communities, for instance, have complained that a state-owned Norwegian wind farm would disrupt reindeer herders' livelihoods (Normann 2021).

Similarly, the transformation to a postcarbon society is going to require enormous amounts of materials — minerals in particular (World Bank 2020; Roberts 2022a, 2022b) — sourced globally through various supply chains, a burden that creates the possibility of setting off a new corporate-imperialist push for vital resources. Outright coups to replace state governments need not be the only method of dispossession associated with resource development. As the history of fossil fuel extraction reminds us, communities on the front lines of energy projects have seen their local environments degraded through pollution, contamination, and so forth, turning them into *sacrifice zones*, places that are destroyed in order to carry out some form of economic activity valued more highly than the life there. Development of raw materials can involve exploitation of workers throughout the global supply and production chain. What have been called conflict minerals are sourced from contexts in which communities are forced to extract resources at literal gunpoint (Bales 2016). It can also involve the destruction of local land bases through mining pollution or overuse of water.

There is some recognition among Green New Deal proponents of this risk (e.g., Aronoff et al. 2019, chap. 4; Neale 2021). One possibility for addressing it is to promote international solidarity between unionized workers in the Global North (who will be essential to building the infrastructure of the Green New Deal) and in the Global South (where rare earth minerals will be mined). Nevertheless, this remains underrecognized, and will be a matter that Green New Deal supporters will need to attend to in coming years.

Consumption and Unsustainability

For neoliberalism, the climate crisis sounds only a minor warning: a problem of market signals. To those thinking along the lines of a Green New Deal, the warning is much louder, sounding the alarm about a form of capitalism that has left large segments of society behind and removed democratic influence over the economy precisely when it was needed to address the climate emergency. But what if the climate crisis is warning of something that goes beyond neoliberal capitalism, and has to do with a specific feature of capitalism itself — perpetual compound growth? Climate change would be just an early horseman of a still unresolved environmental apocalypse.

One potential scenario is that Green New Deal–like programs expand unsustainable levels of consumption, increasing resource use and pollution. One goal of the Green New Deal, after all, is to bring greater shared prosperity to the masses. If the mobilization for World War II provided the model inspiring the Green New Deal's ideas on how to think about a climate response, the postwar period offers a warning about the environmental consequences of mass consumerism. It was in those years that social democratic politics helped establish a strong middle class and saw an astonishing rise in consumption. Given the short time left to win serious action on climate change, it is not surprising to find that proponents of the Green New Deal do not commonly take up this question, though there are some exceptions (e.g., Klein 2019, 264). A Green New Deal might be thought of as a "Last Stimulus" (Aronoff et al. 2019, introduction) to promote the public affluence that would create conditions to allow a break from perpetual growth.

CONCLUSION

The urgency of climate change injected social democracy with a new vitality and relevance in neoliberal times. The Green New Deal represents a reassertion of the belief that if capitalism can be moderated and harnessed by a democratic government, it can promote social justice, support social priorities that unite a sustained and powerful grassroots movement, and provide the means to finally address climate change at the speed required. If the 2020s can become the decade of the Green New Deal, the world might stand a chance of averting some truly unimaginable climate scenarios. Those concerned with climate justice have good reason to organize for and support efforts to win a Green New Deal. That said, it may not address every major concern of climate justice. But it might let forces seeking to win even further progressive social change live to fight another day, to climb a difficult peak they need to before setting off to meet the next. With that, let's turn to what some of these other, still higher peaks might be.

8

DEGROWTH

PATHOLOGICAL CONSUMPTION AND THE DETRITUS OF FAD AND FASHION

THE UNRELENTING NIGHTMARE OF not being able to print our faces on the foamy head of a freshly poured pint of Guinness has at last come to its end thanks to "beverage-top media" technologies. There is finally, too, a mini home refrigerator that will come to you wherever you are in your house so that the fraught and harrowing journey to the kitchen can now be avoided. In that fridge, you might store the wine you can now buy for your pets (so you might drink together in front of the custom bobblehead of your cat or dog that you can also now have made) or the plastic shells shaped like eggs meant for you to pour eggs into after you have cracked their original shells (no more peeling hard-boiled eggs!).

Launched by Guardian columnist George Monbiot (2015), the hashtag #ExtremeCivilization will occasionally mark items like these on Twitter. It gives the platform's users a way to highlight stunningly outlandish instances of the heights of excess our consumerist culture has now achieved. And while it might be tempting to laugh off things as obviously decadent as 3D batter printers or smartphones for dogs, so very many of us have been participants in consumerist excesses, right from a young age. Consumerism has so many ways of enlisting us to buy things we do not (or should not) particularly need. It is probably not an exaggeration to say that the most powerful applications of knowledge generated in the field of human psychology have been in marketing, leveraging all sorts of powerful, deep-seated cognitive and emotional human functions to promote the pathology of consumption. If consumer impulses are not stimulated through fads, nostalgia, or planned obsolescence, there is always the tried-and-true methods of marketing, which turn to anything from preying on our insecurities to selling us lifestyles and values through sophisticated branding.

The frameworks covered so far in this book all agree on one thing: more and perpetual economic growth is good and even necessary. While they might, particularly in these times of inequality, disagree over how the gains from that growth are distributed — whether they accrue to the wealthiest at the top or are shared by the population more generally — those frameworks all converge on a consensus about the desirability of *more growth*. Indeed, there is nothing in the predominant thought or practice of conservatism, libertarianism, neoliberalism, or even social democracy that suggests the economy can reach a point of sufficiency beyond which it need not continue to expand. In each, the model for a decent society is some version of a high-consumption capitalist Western nation, one in which quality of life is ever rising thanks to growing income.

But this "growthist" consensus fails to properly acknowledge something enormous. Our economic system's need to maintain perpetual economic growth is a powerful driver of environmental degradation — including climate change. This is, after all, the other side of the coin of consumerist excess. Just think of the material inputs required to produce the things we buy: plastics, wood, rubber, leather, metals, minerals, paints, fibres, adhesives, inks and dyes, and electronics. And there are also energy inputs in harvesting or synthesizing the raw materials, processing them, manufacturing and assembling them, and transporting them between the different moments of production to storage and to points of sale. Preparing the products for sale also entails packaging (waxes, cardboards, plastics), labelling, tagging, and stickering. These goods also require a system to stimulate their sale, the massive global advertising and marketing industry that itself requires energy and material inputs to plan campaigns and express them over a physical and digital landscape. The stuff of this production — what are sometimes called the material and energy throughputs — must not only come from somewhere (i.e., environmental *sources*) but also ultimately end up somewhere (i.e., environmental *sinks* tasked with absorbing untreated water, nonbiodegradable waste, and, of course, carbon emissions). As that high-consumption model of development becomes increasingly globalized and accelerated, as the throughput is increased, the environmental impact is magnified. The sources get depleted and the sinks get filled to overflowing. And the consequences fall disproportionately on developing countries. This is part of what scholars call "the shadows of consumption" (Dauvergne 2008) or global "ecologically unequal exchange" (Givens, Huang, and Jorgenson 2019; Dorninger et al. 2021; Hickel et al. 2022).

Towards what exactly is the world putting all that energy derived from burning fossil fuels? To what degree are the world's emissions being unleashed to produce, transport, and sell the hollow excesses of #ExtremeCivilization? Would the effort to supply all necessary global energy with zero-carbon sources not be made easier if, globally, we needed less energy?

THE POLITICAL CONDITION OF A "FULL WORLD"

The degrowth framework has emerged as a critical reaction to society's pursuit of unceasing economic growth. From this perspective, there is something profoundly disturbing about a hegemonic mindset that is either unable to recognize the possibility of limits or, even worse, is against limits. Degrowth draws on two sets of critiques. The first is an ecological one, based on that belief that it is impossible to have infinite growth on a finite planet. The second critique is a social one directed at the notion that economic growth is essential for human well-being and will always improve it — that there is no limit on the capacity for human well-being to benefit from ever more consumption of goods and services. In sum, a degrowth perspective sees perpetual growth as not only impossible but unnecessary.

Boundless Desires, Unbounded Economy, Bounded Planet

There is something conceptually unnerving about the pursuit of perpetual economic growth. It is as though the economy is being conceived of as existing on some abstract theoretical plane rather than on a living planet with a limited capacity to provide resources and absorb waste.

Indeed, one of the most common visualizations of the economy — one featuring in so many macroeconomic textbooks — is as a circulation occurring within a closed system (i.e., one that does not interact with anything outside itself). The simplest version depicts, on one side, firms providing households with goods and services, and, on the other side, households providing firms with factors of production (Raworth

2012a). But what is left out of this visualization is that the economy is a subsystem. It is nested within a society, and that society in turn is nested within an environment.

Appreciating this relationship between economy and environment permits important insights. It leads Herman Daly (2005; 2015), the ecological economist whose work on steady-state economies has been influential on the degrowth movement, to rather evocatively call ours a "full world" — that is, a world filled with humanity and its stuff, its buildings, bridges, roads, farms, machines, pollution, and so on.

Just how full has our world grown? There is an abundance of signs that the planet simply cannot sustain a global society aiming to attain Western middle-class lifestyles for all and continuing to grow perpetually even from there. Let's take this moment to peer through some windows onto our current sustainability crisis.

We can start with the image many students are taught in elementary school, the ecological footprint, a metaphor for the impact our way of life has on the environment. Just how big has this footprint become? Every year, the Global Footprint Network calculates Earth Overshoot Day, the day in the year when the demands of the global economy exceed the earth's annual biocapacity, the ability of nature to replace in a year the resources we use and to absorb the waste we emit. In 1970 that day fell on December 29. In the 1980s it came in November. By 2022 it arrived in July. To sustain current global patterns of use, we would need 1.75 earths (Earth Overshoot Day 2022). A European lifestyle for everyone in the world would require three earths, a North American lifestyle five (Global Footprint Network 2022).

Another window. The World Wide Fund for Nature (WWF 2020) Living Planet Index monitors 20,811 populations of 4,392 species of mammals, fish, birds, reptiles, and amphibians, giving a sense of the state of the earth's animal biodiversity. Between 1970 and 2016, the average change across animal populations around the world was a decline of 68 percent. The world's hotspot is South America. There, the change in animal populations averaged a drop of 95 percent in that period, driven by land-use change, climate change, and invasive species.

Yet another window: the mass of the unnatural. One way to get a sense of the scale at which human systems crowd out natural systems across the earth is to estimate the weight of our stuff and compare it against nature's. The estimated weight of domesticated birds (primarily chickens), for instance, is about three times that of all wild birds. Wild mammals (i.e., mammals that are not humans and not livestock) account for just 4 percent of the weight of all mammals on earth. The weight of the human population is ten times that of wild animals. Human-manufactured stuff — our buildings, clothes, electronics, and more — is on the verge of outweighing all of life on earth (Folke et al. 2021).

Still another window. The planetary boundaries framework (Rockström et al. 2009; Steffen et al. 2015) is an important scientific undertaking attempting to define a *safe operating space* for human social development. It identifies several thresholds in the human impact on environmental processes that, if crossed, have the potential to destabilize the entire earth system that has given rise to the conditions of the last roughly twelve thousand years (i.e., the Holocene) — the only conditions we are certain can accommodate current human social arrangements. Two of these planetary thresholds have already been transgressed. The first is the annual rate of species extinction, or change in biosphere integrity. The second, changes in biogeochemical flows, was crossed primarily due to fertilizer use having dangerously altered the way phosphorus and nitrogen are extracted, distributed, and concentrated throughout the world, with potentially dire consequences for aquatic ecosystems functioning. After being used in fertilizers, they concentrate in aquatic systems, setting off a chain of events that deprives the water of oxygen and creates massive dead zones. Two additional thresholds are very close to being crossed. One concerns land-system change, given that just 62 percent of original global forest cover remains. The other, of course, concerns climate change.

These windows on unsustainability are what proponents of degrowth have in mind when they argue that infinite growth is impossible on a finite planet. And the problem of growth is not only in its intended perpetuity, but its intended *rate*. The ideal annual economic growth rate under capitalism is around 3 percent, which means that the scale of the economy doubles every 23.5 years. In other words, the economy of 2043 would need to have grown to twice the size it was in 2020, and that economy of 2043 will itself have to double by somewhere around 2067. Who can conceive of an economy of this scale, never mind one that is also sustainable? The world is big, but not infinite.

The Well-Being–Consumption Paradox

Despite all of this, infinite growth continues to be the order of the day. And so we need to ask, what is the link between our (over)use of the environment and human well-being? In a major 2018 study, O'Neill et al. mapped the countries of the world according to the quality of life they offered and how sustainable their economies were. In general, the better the quality of life in a country, the more environmentally unsustainable it was. As the study's lead author put it,

> Imagine a country that met the basic needs of its citizens — one where everyone could expect to live a long, healthy, happy and prosperous life. Now imagine that same country was able to do this while using natural resources at a level that would be sustainable even if every other country in the world did the same. Such a country does not exist. Nowhere in the world even comes close. (O'Neill 2018)

Are we doomed then to a choice of living in a society that is (a) deprived but sustainable or (b) healthy, happy, and prosperous but unsustainable? Not necessarily. What that bleak global picture does not show is whether the second set of societies might have hit a point where continued growth no longer contributed to health, happiness, and prosperity but continued growing well past it and into the realm of unsustainability.

How can this be true? Doesn't getting more of what we want make us better off? Doesn't it truly and meaningfully improve our lives? Doesn't money actually — contrary to the saying no one is really meant to believe — buy happiness? Proponents of degrowth say no, and point to what is known, variously, as the "well-being–consumption paradox," the "happiness–income paradox," or the "Easterlin paradox (after economist Richard Easterlin, who first wrote about it in the 1970s). It calls attention to a phenomenon in which rising levels of income are associated with rising levels of happiness or sense of well-being — but only up to a certain point. The correlation between them reaches a threshold beyond which the capability of more income to permanently raise happiness is greatly diminished or disappears entirely. That point appears to be somewhere around where a person can comfortably meet essential human physical and psychological needs, sometimes called a *sufficiency threshold*, one that today's developed countries have long ago passed.

What explains this paradox that goes against all capitalist norms we are instilled with? First is a psychological process called hedonic adaptation, which describes the phenomenon of how so many changes in our lives — positive or negative ones — have only temporary effects on our sense of well-being. The duration and intensity of our emotional reaction to these changes tends to fall short of what we might expect when we make our plans of life. It is as though we have a sort of average level of happiness and, though changes in our lives can lead us to deviate from it for a time, we acclimate to them and eventually return to the baseline.

Much of consumption fits into this. Readers have likely had the experience of longing for some good, finally acquiring it — and then no longer finding it particularly special. And this hedonic adaptation effect is not isolated to commodities we consume; it also applies to growing individual income. People believe that their happiness will rise with more money, yet overlook something vital: though greater wealth allows us to meet the consumption goals that were unattainable at lower levels of income, with rising income come rising aspirations. People reach a new, higher material standard of living only to become accustomed to and dissatisfied with it and develop new consumption goals that an even higher level of wealth is required to achieve (Victor 2008, 124–28). We may finally earn enough to afford that BMW we always wanted, but now what about the Bugatti?

In addition to hedonic adaptation, another explanation for the failure of economic growth to promote happiness past a certain level of income has to do with what are called positional goods. We sometimes believe that happiness will come by achieving status over others. One common way to express superior status outwardly is through conspicuous consumption. But as an economy grows and as more people are able to afford positional goods, the goods soon cease to convey unique status.

These effects — a temporary improvement in well-being due to positive changes — are part of what is sometimes called the "hedonic treadmill," a metaphor capturing how, at least with some pursuits of happiness, there is no long-term improvement in our well-being once we achieve them; we simply run in place. We want something, get it, but then grow accustomed to having it, and so seek the next thing we believe will make us happy. *Desire, acquire, acclimate, repeat.*

Meanwhile, the social and environmental consequences of pursuing that growth offers yet another explanation for the well-being –consumption paradox. Daly (2005; 2015) has written about "uneconomic growth," which occurs when the costs of growth exceed its benefits. At this point, instead of accumulating more wealth, we are accumulating "illth," a kind of antiwealth that makes us poorer. Uneconomic growth can manifest in multiple ways — for instance, the attitudes we are compelled to adopt in order to do our part in driving the growth machine — like careerism, hypercompetition, and addictive hyperconsumption — or in the side effects of this lifestyle like loss of leisure, pollution, resource depletion, indebtedness, stress from competition, work precarity, the financial crises caused by reckless strategies to increase profit or resulting austerity policies meant to restore growth's benefits to elites, and climate change.

The implications of this paradox are far-reaching and potentially radical. If squeezing out further economic growth happens at the cost of the free time we would otherwise spend doing what we love with the people we love, or at the cost of our planet's rich biodiversity or of a habitable climate, then is its pursuit not deeply irrational? What is further growth for if it is not increasing well-being — at least in already wealthy countries that have already attained a reasonably high standard of living?

CLIMATE AND DEGROWTH

So how does the matter of growth and degrowth challenge the way we see climate change and justice? For one, the climate crisis calls for an urgent reflection on the role of economic growth in contemporary society, particularly among rich countries where, if the degrowth critique is right, it has ceased to be necessary for human well-being. How has economic growth come to colonize our minds and shape our societies so that we believe it is needed now and forever, and what economic drivers of behaviour and systemic imperatives perpetuate its hegemony (Hamilton 2004, 2010; Hagens 2020; Hickel 2020b, chaps. 1–2; Wiedmann et al. 2020)? What intolerable risks with regards to the climate is the pursuit of growth driving us to pursue (Klein 2010)? Climate justice impels us to ask questions like these to loosen the grip that perpetual growth has on our imaginations and disrupt the sense that it is natural, necessary, inevitable, immutable, healthy, and utterly good.

This disruption is also important because deprioritizing economic growth could permit much more ambitious climate policy. If the growth imperative can be abandoned, it is possible to make much sharper demands under this framework in the face of climate change (Victor 2012; Daly 2014, 87–92; Keyßer and Lenzen 2021). This is because we can ask about the ultimate purpose of an economy. If it is simply to perpetually grow and generate more wealth, as it is under capitalism, then just about any (legal) economic activity that generates monetary value ought to be welcomed. But if we start to inquire into what the human good really is, how an economy can serve that, and when it has reached a point where it provides enough to all, we can also start to inquire into what parts of it are unnecessary and excessive. Just how much of consumption in the developed world is actually improving well-being and how much of it is just to keep us sprinting in place on the hedonic treadmill? As part of averting a climate catastrophe, should we not think twice about emitting carbon just for the commodities of #ExtremeCivilization? Given the outsized role of the richest 10 percent of humanity in emissions, should we not prioritize capping their consumption?

In this spirit, a study in 2020 by Millward-Hopkins et al. estimated how much final energy would be required to provide everyone in the world (at the projected population level for 2050 of ten million) with the necessities for leading a decent life. Remarkably, the authors found it would not take very much at all: 149 exajoules in their estimate, or a reduction of more than 60 percent compared to current consumption. And, the authors noted optimistically, that amount is just a little more than double the energy being provided by non–fossil fuel sources.

Degrowth would potentially involve a reduction in work hours and days as well, which would lower greenhouse gas emissions involved in the various productive processes employing the world's workforce while simultaneously expanding time for leisure. (Work on this matter shows that that leisure time must not be carbon-intensive lest a carbon "rebound effect" kick in [Druckman et al. 2012; Shao and Rodríguez-Labajos 2016].)

Climate responses consistent with this framework also include a re-embracing of the community and the local. A consistent concern over the career of writer and climate campaigner Bill McKibben is the impact that human society at its contemporary scale is having on the earth. We have accumulated, to use his words, too much leverage. With relatively little effort or intention, each of us, particularly in richer countries, can contribute to processes that are changing the earth system. But McKibben (2019a) sees a chance to change this in the response to climate change. Solar energy and the nonviolent climate campaigns to speed its adoption are, as he describes, "technologies of maturity" promoting "economic and political maintenance and contentment" rather than "exhilaration and extension."

McKibben (2010) has also argued that greater localization would reduce the need for transportation of people and goods, reducing in turn material and energy needs. He detects signs of a desire for this kind of shift — or examples of how one can unfold — in farmers' markets and the Transition Town movement, which tend to increase interaction between neighbours, and in local currency projects, which are aimed at localizing the circulation of money. Such a project would be about a transition from larger to smaller systems, as well as a move from highly centralized to decentralized ones. McKibben considers alternatives for meeting energy (i.e., light and heat) and food requirements in decentralized ways. The vast, mechanized, centralized factory farms and plantations made possible by massive fossil fuel–based fertilizer inputs will have to give way to smaller, more independent, and more labour-intensive farms that grow a diversity of crops and that are actually more resilient to climatic extremes. Energy, meanwhile, could come from community-supported energy projects like wind power cooperatives, rooftop solar panels, and small hydro (in subsequent comments, McKibben has rescinded advocacy for woodchip-fuelled boilers). Government subsidies would need to shift away from large centralized projects and towards these kinds of food and energy alternatives in order for them to prove viable. These localized, decentralized communities could potentially prove more resilient against the genuine threats engendered by climate change.

DEGROWING: A RICHER AND MORE JUST LIFE WITH JUST ENOUGH

If its proponents are correct, degrowth requires a much different understanding of the earth, the human good, the economy, and the relationship between them to replace the dominant and deeply anthropocentric understanding now dominant. If material accumulation and hyperconsumption should no longer form the basis for the good life, as it does in capitalism, what should — and in such a way as to step off the pathway of perpetual growth?

Dethroning the Flawed Metric of GDP

The degrowth project seeks new metrics of well-being. The metric used to assess economic growth is GDP, or gross domestic product, which measures the total market value of goods and services within a country, typically by adding up all the consumption, business investment, government spending, and net exports in its economy. Growth in GDP is believed to signal a strong and healthy economy. It indicates that enough of the investments being made throughout that economy are profitable. It suggests that economic activity is dynamic enough that jobs are being created to ensure employment stays high even as the economy sheds work that is no longer in demand, whether because a particular sector has become obsolete, jobs have been offshored, or technology has made operations less labour-intensive. And a high rate of growth suggests that a government can spend more on social programs without raising taxes.

Dividing a country's GDP by its population offers a general sense of the quality of life that should be expected in that economy. High GDP per capita suggests that people have disposable income to spend on goods and services far beyond those required for mere subsistence. They can invest in high-quality ways of servicing their needs, such as good food, nice homes, and private transportation; and they can also engage in extracurriculars that make life more enjoyable — entertainment, culture, hobbies, leisure. And, indeed, countries with higher GDP per capita do in general seem to enjoy a higher quality of life. The 2020 World Happiness Report, for instance, finds that the ten happiest countries from 2017 to 2019 were Finland, Denmark, Switzerland, Iceland, Norway, Netherlands, Sweden, New Zealand, Austria, and Luxembourg — all nations with high per capita incomes (Helliwell et al. 2020).

But the problems with GDP are severalfold, and they are essential to understanding the argument for why a different metric ought to be given priority. First, GDP gives no indication of how income is distributed across an economy. A high and growing GDP per capita can disguise chasmic inequality in which the top echelons hoard most of the wealth and a massive bottom tranche of the population is left in penury. Similarly, it can disguise the social ills that occur due to that inequality (Wilkinson and Pickett 2011).

Second, GDP simply measures the market value of what is produced in an economy; it is unconcerned with whether what it measures is actually desirable — and whether what is desirable is actually measured. There is a large part of what is good or even essential in life that does not register in GDP, while a great deal of "bads" contribute to its growth. The implications for the environment deserve special attention. If the metric of GDP growth is what policy prioritizes, a society ends up valuing the economic activities that drive ecosystem degradation and extinction over leaving those ecosystems as they are. Rendering an ecologically diverse forest into a monocrop for palm oil, for example, is great for national GDP.

The strangeness of GDP is further revealed when compared to metrics used for healthy maintenance. When it comes to, say, human bodies, we have indicators — body temperature, cholesterol, blood pressure, and so on — that warn of ill health as measurements exceed a particular range. GDP is not this type of measure, depicting all and further increase as good. Development economist Kate Raworth (2017, 210) notes a disturbing lack of concern among mainstream economists about the long-term trajectory of perpetual growth. Referring to its absence even from basic economic textbooks, she speculates the eternal upward GDP curve is "too dangerous to draw": in trying to image it, economists are either forced to face the possibility that growth will eventually need to come to an end or conceive of a way for it not to blow well past planetary limits.

A transition towards a new kind of economy starts with challenging the dominant conceptual model of the economy (Raworth 2012b). One possibility is Raworth's (2017) Doughnut Economics. The memorably quirky name is derived from the image she argues ought to replace that dominant mental conception of the economy as being like an unbounded circular flow. It is composed of two concentric circles. The inner circle represents the social floor below which is a state of deprivation that no society should fall into. The outer circle represents the ecological ceiling beyond which we enter a state of environmental unsustainability. The goal of the twenty-first century is to build economies where everyone exists within the doughnut between the two circles, having enough to live full, dignified lives, but doing so within the earth's natural boundaries. There is also by now no shortage of proposed alternative metrics to supplement or replace GDP with indicators measuring, variously, the state of a society's ecological impact, life satisfaction, equality, education levels, infant mortality, life expectancy, and more (Giannetti et al. 2015; Hickel 2020d). In addition to new mental imagery and indicators, the project of dethroning GDP could also entail a new vocabulary of progress and well-being, replacing associations of "more" with "better" (Kallis 2019, 72–73).

Reenvisioning and Reconceiving the Good Life in Postscarcity Economies

The search for an alternative sense of the human good begins with recognizing that many economies have now arrived at a unique historical-economic condition: they have solved what is sometimes called the "economic problem," the need to overcome conditions of scarcity in order to ensure that everyone in a population has enough to live a full and dignified human life (Hamilton 2004). Interestingly, two of liberal capitalism's most influential thinkers, John Stuart Mill [1848] (1988) and John Maynard Keyes ([1931] 2010), anticipated a day when further growth might no longer be necessary or desirable and advanced capitalist economies could settle into a postgrowth condition. Continued economic growth can be justified where it is necessary to lift people out of poverty and deprivation, but for the postscarcity economies of the Global North there is a different policy goal that should take priority: to lead the way in building models for the promotion of that alternative — and ecologically rational — sense of the good.

If there is a concept that captures the essence of the good in a postgrowth world, it is the embracing of the art of life, the ideal that life is for living — and living well by all — rather than for toiling, spending, or consuming. Keynes ([1931] 2010, 331), for instance, imagined it as one where people are taught "to pluck the hour and the day virtuously and well" in a society that had succeeded in shrinking necessary work to fifteen hours a week. Contemporary degrowth writers put an ecological twist on the human capabilities approach to justice reviewed in Chapter 7, which aims to promote human flourishing, or *eudemonism* (Hamilton 2004, 212). For example, Tim Jackson (2009, chap. 9) bases his influential postgrowth vision on an ethos of "flourishing within limits." Latouche (2009) argues for rooting degrowth in human autonomy and conviviality. Robert and Ed Skidelsky (2013) have argued that a good life without growth is found in guarantees of the basic goods of health, security, respect, personality (i.e., ability to pursue one's life plans), harmony with nature, friendship, and leisure. Hickel (2020b, chap. 5) writes of a life characterized by radical abundance, which, contrary to growthist belief, can only flow freely in a degrowth society that has left behind the artificial scarcity in free time, public goods, employment, and more created under capitalism.

Antiwork

Another window into the possibility for expanded freedom offered by degrowth is the focus on work and the place it holds in the popular conception of what well-adjusted adulthood is. The expectation throughout virtually the entirety of the capitalist world is for people to attain a full-time job (or multiple simultaneously held jobs) occupying somewhere around forty hours (often much more) per week from one's early to midtwenties until retirement in one's sixties or seventies. This is an *enormous* amount of our lives given to work, which is all well and good if we enjoy what we do (and feel we do so freely). But so many do not. Advocating for a reduction of working hours in exchange for more leisure time is one of the ways a degrowth movement could gain some mainstream traction. A society not compelled by growth could make more rational decisions about the work that needs to be done and share it equitably, leaving more time outside of work for people to engage in building and nurturing relationships or taking on their own creative or inquisitive projects as part of developing their capabilities.

Recovering — and Rediscovering — the Shared

Fears about a diminishing quality of life due to having access to fewer goods in a postgrowth world can be addressed by expanding what is socially shared. A society cannot live within ecological limits if all its members pursue individual private luxury, but that does not mean they cannot enjoy social luxury. Investments in high-quality public goods and amenities — parks, galleries, sports centres, transportation, and more — intended for all are part of the ethos George Monbiot (2017a) has described as "private sufficiency and public luxury." A postgrowth society could also revitalize the commons, which, broadly speaking, refers to essential resources (e.g., forests, water, or even forms of traditional knowledge) that are collectively protected and governed by the communities benefiting from them — rather than the state or private owners — to ensure their sustainable and equitable use. If care is taken to pursue it in the context of Indigenous and ecofeminist concerns about decolonization and gendered power relations, commoning (or recommoning) of privatized or state-owned resources can offer an opportunity to think about equitable and democratic economic self-governance (Perkins 2019).

Life at a Human Scale

Degrowth advocates also urge a relocalization of economies, arguing that, from an ecological perspective, the global trading system is irrational. For example, vast energy and material inputs are involved in making particular foodstuffs available throughout the world and throughout the entirety of the year. Turning to local agriculture could be a way of addressing this irrationality, as many "locavores" would no doubt be proud to tell us. And the same might be said for other sectors of the economy that have been stretched out across the whole of the world in the search for lowest-cost production.

But in addition to reducing the planetary space across which production for rich-world consumption unfolds, the scale of everyday life could shrink too. We might imagine looking at our world today through the eyes of someone in a future where a degrowth consciousness had replaced our growthist one. Might they see large houses with rooms or even whole floors that go unused except to store items hoarded over a lifetime of consumption as a strange compulsion from a time when it was assumed the world was limitless? Might they see an irrationality in the arrangement of urban space and ask why neighbourhoods were not organized to be navigated without cars? (Why *did* people insist on driving absolutely everywhere?) What strangeness would they see in the tendency of growthist societies to reduplicate appliances and tools and so forth that mostly sit idle? Why not cook together in the kitchen commons? Why not just use the community laundromat to wash clothes? Or visit the neighbourhood all-libraries to find new clothes or to borrow that drill set or those cooking dishes needed so rarely through the year? Why keep throwing away and replacing broken things when the community repair shop can get them working again?

Minimizing the size of the economy and eliminating its drive to tear down ever more barriers to growth also permits political actors in rich countries to argue in stronger terms to leave climate and ecological space for poorer countries to grow their economies in pursuit of development. So, too, would it arrest the colonial drive to exploit traditional Indigenous territories and upend their ways of life.

THROUGH A CLIMATE JUSTICE LENS

Revisiting our questions of climate justice, there is much to embrace in degrowth. It permits its proponents to shape an ambitious approach to climate change by taking into consideration what an economy should actually be for and when it has achieved sufficiency; additional, extraneous economic activity — and the energy required for it — can be downscaled. But this section highlights two concerns about degrowth. First, of the frameworks we have seen so far, degrowth has the strongest potential to leave ecological space for nonhuman life, but does it emphasize this enough? A second has to do with whether degrowth can realistically offer a means of addressing the climate crisis — or, for that matter, even come into power — while also preserving desirable elements of contemporary liberal societies. If not, it might be using up scholars' and activists' time that would be better put towards different projects, either less or more radical, or perhaps it should spur thinkers to dedicate more effort to how the benefits of degrowth can be brought into the real world. We explore these matters by looking at degrowth from liberal and socialist viewpoints.

Accommodating the Nonhuman

In adjusting the size of our economy to levels compatible with environmental sustainability, just what should sustainability mean? What does it really mean to pass environmental limits? Are they only passed once we begin to affect human well-being by degrading or losing environmental services that humans value or even require? Or should the well-being of nonhuman nature count for something? The degrowth literature tends to primarily be concerned with environmental sustainability for the sake of human well-being, or at least foregrounds its concern about environmental limits with reference to how transgressing them is bad for humanity.

As such, its sustainability limits might accommodate animal population decline, species extinction, ecosystem destruction, and biodiversity loss only to the degree that the human good is not negatively impacted.

But, to bring this back to the climate crisis, if our concerns are extended outward to also include nonhuman life, the moral case for taking major climate action is strengthened because it expands the communities on the front lines of the crisis. Democratic deliberation could be expanded and enhanced to more fully include the concerns of nonhumans, including both animals and ecosystems — an *ecological democracy* or *biocracy* (Ball 2006). Part of fostering and maintaining a degrowth ethic would involve seeing nonhuman living entities as subjects with whom we must enter into nonexploitative, nonextractive, reciprocal relationships (Hickel 2020b, chap. 6).

Degrowth vs. Liberalism

As degrowth attracts more attention, it will predictably attract more scrutiny from those subscribing to neoliberalism or techno-optimism — and indeed that has already begun (Barthold 2020; Smith 2021). But there have been critiques of degrowth from even the more progressive Green New Deal camp. This might be surprising given that proponents of the Green New Deal share a lot of values and goals with proponents of degrowth, both seeking to wrest society from the grips of free-market economic orthodoxy and create an economy based on justice and sustainability in the context of an ecological crisis. Indeed, there are some who believe that the left wings of both camps overlap enough on these sorts of values and goals that the two projects might be fused as one (e.g., Dale 2019; Cox 2020; Mastini, Kallis, and Hickel 2021).

That said, some supporters of a Green New Deal would argue that degrowth offers few real prospects for addressing climate change. Economist Robert Pollin (2018; 2019), who has written extensively about the Green New Deal, is a case in point. While sharing "virtually all the values and concerns of degrowth advocates" with respect to the environmental damages driven by growth, wasteful production, the failure of growth to address inequality, and the problems with the use of GDP, Pollin (2020b) observes that "degrowth is not a solution, just in terms of simple mathematics. Right now the globe generates about 33 billion tons of carbon dioxide emissions. Let's say we cut global GDP by 10 percent, which would be a bigger depression than the 1930s. What happens? We cut emissions by 10 percent, from 33 billion tons to 30 billion tons. It's no solution at all." Pollin here appears to be conceiving of degrowth as a steady decline across the board. But a degrowth program for climate change need not occur in that way any more than an income tax needs to be leveraged at a flat instead of progressive rate. It could start by targeting luxury hyperemissions, for example, the kind generated by that richest 10 percent who are responsible for half of 1990–2015 emissions.

A more difficult problem that both neoliberals and social democrats might see in degrowth is the matter of whether shifting to a degrowth society would risk losing some of the positive dimensions of liberalism that were won under conditions of growth — like democracy, freedom, cosmopolitanism, scientific progress, and even individualism (Dobson 2013). There is, potentially, a too-ready assumption among degrowth thinkers that these goods of liberalism would survive, and a failure to appreciate there might be a "thermodynamic price tag" (Quilley 2013) on the social conditions required for them to flourish.

Love it or hate it, growth is hegemonic. It forms part of people's expectations for their economies. The idea that economic growth equates with well-being is hard to shake, after all. And the thought of facing economic contraction is a painful and fearful one for most people. There is even a kind of truth behind these concerns. When GDP is contracting unintentionally and out of control, hard times befall people. GDP contraction is a sign that something in the system has gone wrong that is making spending through the economy fall. The demand for a host of products and services has dissipated, and with it the need for jobs. The COVID-19 pandemic was a tragic reminder for many of the unemployment, hardship, and poverty that can occur when the economy cannot function at its fullest. (Degrowth advocate Jason Hickel [2020a] published a Twitter thread specifying that the unplanned recession caused by COVID-19 was *not* an instance of degrowth.) Given the short time left to decarbonize in order to avoid surpassing global average warming of 1.5°C in a decade that began with the world seeking to deal with the economic recession triggered by the pandemic, questioning the need to grow seems unlikely to get very far. So what viability does a degrowth project hold in the real world? Could pursuing its critique detract from political efforts to rally sustained support for a Green New Deal?

Degrowth vs. Socialism

Compared to social democratic proponents of a Green New Deal, socialists are generally more aware of the degrowth critique — both frameworks are, after all, contending for how to define a postcapitalist world — and so we can find among socialists a diversity of attempts to wrestle with it. For one, there is a line of socialist thought (sometimes called *accelerationism*) that rejects living according to limits and seeks to usher in a highly automated postcapitalist economic system of sheer abundance (e.g., Phillips 2015; Bastani 2019). From this perspective, degrowth can be seen as unnecessarily circumscribing the full human potential that can be unlocked under a pro-growth socialism, clipping off our ambitions right as our technological capacities have never been greater. But as the next chapter suggests, this perspective is out of step with much contemporary socialist thinking on the environment: ecosocialist analyses do tend to imply limits to the size of an economy.

So another more serious line of criticism from socialists — and one that might be similar to that of Green New Deal proponents — is this: the degrowth project's plan to realize itself seems undercooked. Though there are plenty of policy proposals (e.g., Cosme et al. 2017; Hickel 2020b, chap. 5), the pathway to power, to making degrowth real in the world, needs tending. Socialism has a long tradition of theorizing the roads to a postcapitalist world — whether by taking power through parliamentary means, reforms, or revolution, or even eschewing hierarchical power all together — and the type of political agents and widespread consciousness required to take that path. It has produced generations' worth of analysis on the successes and failures of attempted socialist revolutions and reforms. There is not yet a comparable tradition in the degrowth line of thought (but see Kallis et al. 2020, chap. 5; Soper 2020, chap. 7). Focusing on an abstract property of capitalism — growth — creates an ambiguous relationship with capitalism itself. Some versions of degrowth hold left-wing or radical anticapitalist implications (Latouche 2012; Muraca 2013); other visions of it are achieved within existing institutional structures through reforms (Jackson 2009, 202; Ott 2012). Socialists might insist that it matters very much whether a degrowth society retains capitalism's central institutions — such as corporate structures, undemocratic employment relations, the market (whether for goods and services or labour) — because those institutions hold implications for the degree to which that society has liberatory potential and which segments of the population it would benefit.

If growth must be eliminated, what appeal does this framework have in a world where capitalism rules and has grown hegemonic? What appeal would it have with those in power? What appeal does it have with the working masses who would likely understand degrowth as a reduction in their quality of life? In the absence of convincing answers to questions like these, ecosocialists would find it difficult to see opposition to growth as a solid basis on which to build a movement strong enough to undo such a core concern of the economic and political system.

CONCLUSION

Even if a degrowth pathway to power remains undertheorized, its foundational arguments and outlooks are hard to dismiss. There is something disturbingly wrongheaded about seeking perpetual growth for its own sake, assuming it necessarily improves the social good and that it is somehow possible on a finite planet. Once the idea of degrowth takes root in the mind, it tends to make much of our contemporary economy seem absurd. The products of #ExtremeCivilization become less like pleasing novelties serving to remind us of the taken-for-granted pulses of innovative dynamism essential to some ideal economic system and more like signs of mania, a pathology that is costing us a safe climate. It prompts us to abandon the easy pursuit of well-being through the accumulation of *more* for a deeper pursuit of it through mastery of *the art of life*.

9

ECOSOCIALISM

ECOSOCIALIST IDEOLOGY

Ecosocialism adapts a long tradition of radical social critiques of capitalism — from Marxism and anarchism on the far left, to democratic socialism just left of the social democratic centre — and adds to it an ecological critique of capitalism focusing on how the system drives a series of environmental crises, climate change being foremost among them.

To understand it, let's begin with that tradition of social critique. Over its history, socialism has encompassed a rich tradition of thought and debate about life in — and pathways to — a postcapitalist world (Draper 1966; Hahnel and Wright 2016; Wright 2019, chap. 3), and readers just learning about it should bear this diversity in mind. But one useful starting point for understanding the ideology might be this. Socialism is founded on a sense of equality that is more radical than any of the frameworks we have seen thus far. To a socialist, an equal society is just that: equal. No person should experience a substantially better quality of life than any other person. Everyone should experience the same, high quality of life thanks to equal access to the means for leading a flourishing human existence. And this vision of equality extends to humanity on the global scale.

This starting point explains a number of important socialist characteristics. For one, it drives an intense opposition to any and all forms of domination and oppression. Socialists might therefore agree with progressive liberals on the need to eradicate the various *isms* and *phobias* from society — racism, ageism, sexism, heterosexism, ableism, homophobia, Islamophobia, antisemitism, and so forth. These are, after all, instances of unjustifiable power relations that prevent people from experiencing the world equally and as fully human. But to a socialist, the equality offered in liberal societies is necessarily incomplete; socialism's search for sources of inequality and oppression thus tends to extend much further. To appreciate this, consider Noam Chomsky's (2013) description of libertarian socialism (or anarchism) as

> a tendency that is suspicious and skeptical of domination, authority, and hierarchy. It seeks structures of hierarchy and domination in human life over the whole range, extending from, say, patriarchal families to, say, imperial systems, and it asks whether those systems are justified. It assumes that the burden of proof for anyone in a position of power and authority lies on them. Their authority is not self-justifying. They have to give a reason for it, a justification. And if they can't justify that authority and power and control, which is the usual case, then the authority ought to be dismantled and replaced by something more free and just.

This tendency has, over the history of socialism, called into question an array of human social arrangements, from religion to the state. But the old, implacable enemy that all forms of socialism have always set themselves against is capitalism. For one, capitalism drives unacceptable economic inequality. This is seen in the sharp disparities between the wealthiest 1 percent and the rest in neoliberal times. Socialists and social democrats would agree on this point. Where they diverge is on the matter of retaining capitalism itself. Recall how social democrats attempt to harness capitalism to promote the social good, equality being promoted by redistributing the surplus generated by capitalism to ensure an extensive set of rights, one creating a high floor that no one in a society falls below. A socialist finds this unsatisfactory: not only would a social democratic society still tolerate rather large inequalities so long as that floor was being maintained, but the inequalities that occur under any form of capitalism are always of an irredeemably intolerable sort.

Capitalist Oppression, Exploitation, and Appropriation

This is because there can be no capitalist societies that are free of oppressive relationships. Capitalism necessarily requires a class division between owners of land and productive property, on one side, and workers, on the other. Under this arrangement, workers have no ability to be fully self-sufficient; virtually everything required to produce the goods and services needed to live is held as the private property of the owning class, who produce and distribute goods and services for monetary profit alone. Workers therefore have no choice but to exchange their labour power for money. Once employed, workers are reduced to cogs or tools whose sole function is to perform tasks that promote the enterprise's profitability. In the workplace, they are subject to total control; they have no real say, particularly in the absence of unions, about the number of hours they work and their renumeration, how much vacation they receive, what they are to wear at work, which tasks they perform, which coworkers they have, who they are to report to, and so on. One result of such a society, socialists argue, is that human beings experience a profound sense of disconnection or dissociation from their work as its conditions unfold in ways removed from their control. The consequences extend beyond the workplace. What scraps of free time people have outside of the working day are not enough, being occupied by the need to recover psychologically, emotionally, and physically from long hours performing functions determined by supervisors and bosses. One of the easiest things to do in those moments is to consume or "veg out" in front of a screen.

Workers are also monitored closely. (A reminder during the COVID pandemic came in the form of "tattleware" apps that employers required workers to install on personal devices to monitor their productivity.) Workers agitating for their workplaces to take stronger action on labour and social issues do so at great risk, making themselves vulnerable to firings that leave them without means to access income necessary for survival under capitalism. Even in off-hours, outside the workplace in their private lives, workers have to be wary of what they say and do. We may not feel badly when someone gets fired after a video clip of them indulging in a racist rant in their private life is discovered online. But it's this very same capitalist power that leads people to fear they will be fired for taking a public stance on *any* political issue, a fear experienced by some participants in the 2020 Black Lives Matter protests, for instance (Hess 2020). It creates a chilling effect. Our lives, even outside of work, are ruled by private tyrannies.

This relationship between workers and owners of capital is one that socialists find not only despotic, but inherently exploitative. This is because a portion of workers' labour always goes effectively unpaid. In standard Marxist analysis, this is how capitalists generate profit. Workers, if it were up to them, would generally be satisfied working hours that earn them enough pay to carry on a decent existence. They would happily provide their labour power in exchange for the equivalent of how much value their work creates (and use the money they earn to buy the commodities they need). Let us imagine it takes six hours to earn that amount. Under capitalism, however, this would be too short of a working day: the owners have not yet made their profit. Workers therefore must be made to labour for longer in order to generate value that the owners can then appropriate for themselves and put towards investment or luxury. Just how much longer — how many hours of effectively unpaid work — has to do with what capitalists can get away with. The worker, who would gladly accept a shorter working day, has no ability to say, "The six hours of work I have done is equivalent in value to what you pay me and it will be enough to meet my needs, so I'm going to head home now." This is not an option under the capitalist bargain; they must work for the hours they are told to, including those during which they generate value they will not see a penny of — or else have no work at all. Over history, workers have been moulded or disciplined in a variety of ways to accept this bargain.

Capitalism seizes for the owner class wealth that is generated collectively. This is true of when technologies that were developed through public funding in their early stages become privatized; the riskiest part of the development is socialized while the benefits are privatized. It is true, too, of when owners appropriate for themselves wealth in the form of rents or property values that have risen not because of anything the owners have done, but as a result of nearby vibrant community life or publicly provided transit infrastructure or parks.

While socialists might approve of social democrats' efforts to take portions of the economy (typically health care and education, sometimes more) out of private hands and to institute living-wage regulations, they insist it does not go far enough. Even if workers are able to win gains like these, it is not long before capitalists find ways to reach their hands back into workers' wallets, whether through rent or mortgage costs (driven up by housing market conditions shaped by speculative investment), oligarchic merchants (e.g., telephone or internet service providers), interest on credit cards or money loans, or increasing privatization of once publicly provided or subsidized education, health care, or utilities (Harvey 2014, 67).

Capitalism vs. Democracy and Self-Determination

Socialists would add that capitalism is not just exploitative, but also fundamentally undemocratic. For socialists, a truly democratic society is one in which people are able to participate collectively and equally in making not just political but also economic choices. But by placing so much of the economy in private hands, capitalism removes the major economic decisions about investment, production, distribution, and so on from the public; what is produced, how, by whom, and for what purpose is left up to those whose overriding concern is to make profit. The defence often put forward by capitalist apologists is that, actually, the public does participate in these decisions through the free market, that we "vote with our wallets," which ignores the fact that some wallets are much bigger than others, and that there are some options that capitalists simply never offer for sale. The market may give us plenty of choice as individuals about which personal automobile we wish to buy, but it does not give us the choice as a society to establish free, efficient, safe, and comfortable public transit.

As wealth gets concentrated in private hands, so does political power. Wealthy donors and corporations can invest in campaigns of politicians promising to deregulate or cut taxes. They can hire lobbyists to pressure political decision-makers. Corporations can threaten to leave an area and lay off workers en masse — a major blow to any working-class community — as a way of discouraging or punishing policies that cut into profits. Legislation gets rewritten in the corporate image.

Capitalist societies do not self-determine. Rather, the system shapes society in its image. The very rhythm of life takes on a capitalist tempo. Cities wake and commute and work and return home en masse. Our desires and wants are elicited and nudged and shaped by marketing. Our public space is inundated with endless, pummelling waves of advertising. The spatial arrangement of our built geographies — neighbourhoods and suburbs, industrial corridors, roads, telecommunications, rail systems — reflect the power and interests of capital. Even at the individual level, our very plans of life must take into account the conditions that capitalism creates. If we pursue what we are truly passionate about, will capitalism value our work enough so that we can make ends meet? Or will that passion have to be reduced to a hobby (or perhaps merely a plan to pursue that hobby) because we have to instead turn ourselves into something that can excel in the marketplace? Even the limits of acceptable thought on a variety of political issues are established by an agenda-setting, for-profit corporate news media that decide which issues to cover, how frequently, through what ideological frames, and with which experts and actors spoken to. (Critics often point out how little climate change gets covered as a result [e.g., MacDonald 2021].)

Imperialism and Dispossession

Capitalism draws the larger world under its aegis. In the endless hunt for ever more profit and growth, it always sets its eyes on new territories, lands, resources, and markets. Capitalism expropriates and imposes private property regimes over traditional land and resource management systems (which had often been organized in forms of commonly held and managed property, also called the commons) as well as over public property, assets, and services in order to subject them to the logic of profit maximization. This key historical strategy for growth is what the Marxist geographer David Harvey (2003; 2009) famously calls "accumulation by dispossession," where wealth is appropriated and concentrated through an array of predatory processes used to strip away, seize, or close off assets from the communities that previously held them, and to repurpose them for profit generation. These processes include

> the commodification and privatization of land and labor power, and the forceful expulsion of peasant populations from the land (as in Mexico and India in recent times); conversion of various forms of property rights (common, collective, state) into exclusive private property rights; suppression of rights to the commons; suppression of alternative (indigenous) forms of production and consumption; appropriation of assets (including natural resources); the slave trade (which continues today, particularly in the sex industry); usury; and, most devastating of all, the use of the credit system and debt entrapment to acquire the assets of others, most dramatically represented by the mortgage foreclosures that swept through the United States housing market beginning in 2006. (Harvey 2009, 68)

In the process, as communities are dispossessed of alternative means of subsistence or work, they are pressed into the sphere of exploitative capitalist class relations.

Capitalism adopts its most gruesome and horrific forms in pursuit of this dispossession, forms that historically included racial slavery, settler colonialism, and global imperialism. Neoimperialist endeavours updated these forms for the twentieth century and early twenty-first century, using a variety of means to topple democratically elected governments insufficiently friendly to the geopolitical aims of imperial powers (Chomsky 2003; Blum 2004).

This standard socialist critique of capitalism's effect on society has been made more urgent by the appreciation of the role capitalist logic has played in setting off the environmental crisis. Agreeing with the degrowth framework, ecosocialists see consumption-fuelled economic growth as a powerful driver of environmental destruction. Capitalism's growth imperative not only continually increases the quantity of a given set of commodities to be sold, thereby increasing material and energy use, but also expands the very scope of what can be commodified. Increasing portions of the natural world (and even aspects of it like environmental services) can become subject to the sphere and logic of the capitalist market, which recognizes the value of the environment only to the extent that it is commodifiable, ignoring nonmarket values like biodiversity or sustainability. This dynamic makes it incapable of respecting limits — social or natural — and is intensified by capitalism's need to accumulate profit in the short term (Kovel 2002; Magdoff and Foster 2011, 37–60).

But ecosocialism tends to highlight additional environmentally destructive dimensions of capitalism sometimes underemphasized in the degrowth literature. For one, the framework tends to hold a deeper and different analysis of the power inhering in the state-capital relationship than in other frameworks, understanding the state as having been captured to a very large degree by the capitalist class, and having taken on the primary function of actively driving capital accumulation, all exacerbated by the ascendance of free-market neoliberal ideology. Capital thereby influences the state in multiple ways to increase opportunities for profit at the expense of the environment: it presses decision-makers to rescind hard-won environmental regulations or block new ones; it influences the negotiation of trade agreements to limit permissible environmental protections and allow the penalization of government decisions to cancel environmentally destructive projects; and, through privatization of state services, it turns over to market actors decisions on important services with environmental implications — such as energy generation — that would better be left in the hands of communities or the public. With state assent, capitalism also sites the most destructive extractive or productive processes near marginalized communities (in both the Global North and South), creating *sacrifice zones*, places where people are abandoned to suffer pollution in the name of profit. All of this has left a legacy of regional poverty and underdevelopment, sharp global inequality, and ruined ecosystems.

Socialists have always seen a certain irrationality in capitalism that needs to be transcended. In precapitalist societies, production was mostly conducted for use. But under capitalism, the basis of the economic system is only incidentally related to satisfying human needs; its driving imperative is to produce commodities for profitable exchange. An ecosocialist vision of the good society seeks, instead, to meet human needs — including the need to access a safe and functioning environment — by liberating or shielding more of social life from the dictates of capitalist market rationality (Gorz 2012).

That will require a deepening and extension of democracy beyond the one found under the liberal order. Under social democracy, equality is promoted by ensuring an extensive set of rights that creates a high floor that no one in a society falls below. But a social democratic society could still tolerate rather large differences in economic status so long as that floor was being met, meaning it would still retain indefensible inequality. Neither could any suite of reforms long contain capitalism's urge to shape society according to its logic, and none will purge society of the oppressive relationship between owners and workers. And so, in addition to resisting capitalism, socialists have always imagined alternative economic arrangements, whether by specifying the basic principles and essential building blocks for their development (e.g., Wright 2010; 2019, chaps. 2 and 4) or providing blueprints for the way a socialist system could work (e.g., Michael Albert 2003) so the *demos* is an active participant in economic decision-making.

ECOSOCIALIST CLIMATE RESPONSES

The climate crisis exemplifies so many concerns of the ecoleft. Capitalist growth created the impetus to seek out and exploit the cheap energy provided by fossil fuels to power the various processes involved in the global production of commodities. The same need for growth and profit has led capital to corrupt democracies and influence state policies in a number of ways to ease regulations hampering fossil fuel extraction and to block climate policy domestically and internationally. Above all of that, the capital class has pushed for lowering taxes on the private sector, privatizing public services, and constraining democratic economic planning — precisely when the public sector is most needed to address climate change. Capital's reckless development of dirty energy has left communities suffering the colonial legacies of poverty and political marginalization on the front lines of both fossil fuel extraction and climate impacts.

The phenomenon of climate change denialism is entirely in line with the capitalist drive to prevent all barriers to accumulation through whatever means available, including corruption of democracy through propaganda campaigns funded by industry. When liberals celebrate capitalism as a source of innovation and production for clean energy technologies, they ignore the fact that the main effect of capitalism so far on climate change politics has been denial, delay, and incrementalism. Not only does capitalism drive climate change and prevent policies to address it, but it is also a dangerous and unethical system through which to respond to it. One dangerous tendency is to propose false solutions to the crisis, which involve responses that are less concerned with reducing emissions or keeping fossil fuels in the ground than they are in protecting processes of capitalist accumulation throughout the transition to a postcarbon economy. Critics count among these false solutions carbon-pricing mechanisms and even carbon taxes (e.g., Gilbertson 2017), but the example that has drawn the most ire from socialists are cap-and-trade schemes. Ecosocialists argue that these schemes have done little to reduce emissions. Rather, they privatize a vital global commons essential for life — the atmosphere's ability to safely absorb greenhouse gas emissions — and operate on the undemocratic assumption that governments have a right to turn control of part of the climate system over to corporations. Where carbon offsets are introduced into emissions trading schemes, they add the further danger of colonial land-grabbing.

The climate crisis is thus a signal that there must be a radical engagement with capitalism (and not merely aspects of capitalism like market failures or perpetual growth) as part of the response to it. But there is a challenge to discussing an ecosocialist climate response. Many socialists have recognized that the short time left to address climate change means that a response must occur from *within* capitalism. For instance, Ian Angus (2016, 214), author and editor of prominent online ecosocialist journal *Climate and Capitalism*, writes that a successful ecosocialist movement would need to be broad based and welcoming of a diversity of ideas, and warns against the rejection of programs because they are not sufficiently radical. As the editorial board of Britain-based anticapitalist network Socialist Resistance (2020) write, "We are not opposed to revolutionary change — quite the reverse. But to gamble the future of the planet on such an unlikely scenario as global revolution within 10 years is reckless in the extreme." Or, as Noam Chomsky (2020) put it, "We should recognize that if global warming is an automatic consequence of capitalism, we might as well say goodbye to each other. I would like to overcome capitalism, but it's not in the relevant time scale. Global warming basically has to be taken care of within the framework of existing institutions, modifying them as necessary. That's the problem we face."

Bestriding the ideological divide between social democracy and ecosocialism, Canadian journalist, activist, and author Naomi Klein has offered a sense of what that program might look like. Her 2014 bestseller *This Changes Everything* laid out a vision for a climate response that closely anticipated the stronger versions of a Green New Deal. For her, the kind of radical engagement with capitalism demanded by the short window remaining to respond to climate change is to be directed primarily at reversing the neoliberal logic that has dominated economic policy for the past forty years (Klein 2014b). It is a response centring on a massively enlarged role for public planning, a key dimension of which is rapidly phasing out fossil fuel resource development. Services that were privatized or defunded under free-market orthodoxy will need to be returned to the public sphere and expanded to rapidly lower emissions and respond to climate disasters. They include accessible and efficient public transit, building-retrofit programs, climate-smart urban planning, emergency and disaster response, and even nonprofit insurance. Most crucially, energy services will need to be brought under public control (and preferably run as cooperatives or commons) because privatized, for-profit utilities cannot be trusted to make the kind of immediate and massive investment in renewables required to rapidly bring down carbon emissions where these investments do not guarantee them short-term profits. But to undertake this project, the underfunded public sector will need to be replenished, and that will require getting the worst polluters to pay their due through measures like steep carbon prices, higher royalty rates on fossil fuel companies, an end to fossil fuel subsidies and military investment, and a billionaire's tax. For measures like these to take hold, a battle of ideas must be won that delegitimizes the ideological basis of free-market capitalism and replaces it with one that rebuilds "the very idea of the collective, the communal, the commons, the civil, and the civic" and restores "the right of citizens to democratically determine what kind of economy they need" (Klein 2014a, 125, 460). A key strategy will be to win policy battles, like those around basic income guarantees, that assert an alternative and more just worldview than the one offered under neoliberalism.

Programs like Klein's — and similar ones advocating for a radical Green New Deal–type economic mobilization (e.g., DSA Ecosocialist Working Group 2019; Ministry of Just Transition Collective 2022) — strain at the outer edges of social democracy while laying the groundwork for efforts further down the road to transform society in more radical ways. This is where the ecosocialist framework brings to the climate fight its most unyieldingly ambitious edge, tapping into an inexhaustible source of refusal to compromise unnecessarily with capitalism. That eye on prospects for expanding the horizons of emancipatory transformation is why ecosocialists tend to emphasize the decommodification aspects of the Green New Deal, seeking to remove housing, energy, jobs, and education (and, in the United States, health care) from the logic of for-profit market exchange, whether in the creation of new, public enterprises or nationalization of existing industries. There is no reason to stop at this point, where social democrats might be satisfied that enough has been accomplished; ecosocialists would press for more, generally arguing that winning more of these policy battles and changing material conditions would create even further support for ecosocialist transformations.

What might occur in this "higher and more radical stage" (Socialist Resistance 2020) of struggles for liberatory and ecological social transformation? The utopian imagination grows in its ambitions. It might entail the decommodification of CDR technologies by making them publicly owned, nationalizing oil and gas companies to turn their functionalities towards carbon capture while giving their workers the option to stay on, or taking lands used for the meat industry to allow natural forest regeneration (Malm and Carton 2021) as mentioned in the geoengineering chapter. Or it might involve the socialization of finance and of corporate enterprises, and the recommoning of nature, technology, media, and more (Lawrence and Laybourn-Langton 2021). The eroding of borders to welcome people compelled to move because of climate change is another such possibility.

From there are all sorts of models that ecosocialists have imagined for much different societies than those that would occur under liberal ideologies (e.g., see Bookchin 1989; Magdoff 2014), including for the governance of the earth system freed from capitalism (Michael J. Albert 2020). But for now, the fight will need to remain focused on immediately winnable struggles for reform that can open up space for more radical ideas to take root.

A FINER LENS

Even though a full transition to a socialist society in time to avert the climate crisis is hard to imagine, the tradition of radical left-wing critique can add some deeper poignance to critical issues or highlight matters that the previous frameworks miss or undertheorize. This section draws attention to the richness that ecosocialist critiques can add to climate justice analyses.

History and Reparations for the Crimes of Capitalism

Andreas Malm (2016b) proposes a few reasons that historians ought to study the history behind contemporary climate change (rather than their more common practice of investigating how brief changes in climate have affected human history). For one, it would explain the social and economic forces that lay behind the initiation, spread, and intensification of the fossil fuel economy that has propelled us into this crisis. Naming these active forces might weaken the sense that they are historically natural or inevitable and make them easier to confront. Such a history could also identify more sustainable modes of life passed over or extinguished in the sweep of the fossil economy. And it might also change the way we understand historical responsibility for climate change — and who owes what today as a result.

Malm argues that the historical force responsible was and remains capitalism. In *Fossil Capitalism*, Malm (2016b, 241) details how the switch from water power offered by river systems to fossil fuels in the early phases of the English Industrial Revolution had less to do with the latter being more powerful, abundant, or cheaper — as the conventional narrative goes — and more to do with the control they gave to capitalists over space and labour: "The fossil economy is the singular offspring of a distinctly capitalist economy, on whose wings it spread to other parts of the globe." And that spread was not peaceful. "Concocted in a constricted core," Malm (2016b, 236) writes, "steam power was explicitly conceived as a weapon to augment the power over the peripheries, haul in the products of all continents, dispatch manufactured goods in return, and ensure military superiority all along the way, in a sort of fossil-imperial metabolism that undergirded the post-1825 development of empire."

Using this kind of view on history, we can scour the past for histories of expulsions and coups orchestrated by imperial powers in the name of securing control over fossil fuel resources or creating the infrastructure to transport it. We can also better appreciate that what was established over the course of this history was a global division of consumption and labour with the Global North accruing the benefits. Brand and Wissen (2021) call this "the imperial mode of living." Today it is not just lands that have been colonized, but, as many in the climate justice movement remind us (Warlenius 2018), the atmosphere as well, due to the carbon emissions of mostly wealthy advanced capitalist nations and of individuals and the ongoing effort to sell property rights to the atmosphere through cap-and-trade schemes.

A first set of responsibilities concerns obligations of governments in the Global North to communities living with the legacies of colonialism in the Global South, understood as a climate debt. These responsibilities would require developed countries to commit to drastically cut emissions, to transfer expensive clean energy technologies (without patents) to the countries of the South to allow them to leapfrog dirtier development paths, to pay the South's climate adaptation costs, to change immigration laws to take in climate refugees, and to alter their own historically rapacious economic patterns (World People's Conference on Climate Change and the Rights of Mother Earth 2010a; Klein 2009, 2014a).

Ecosocialist authors also highlight the importance of linking anti-capitalist climate struggles with Indigenous, environmental justice, and labour movements. Indigenous communities and their traditional territories are often on the front lines of fossil fuel extraction, pipeline transportation, and refinement (Indigenous Environmental Network and Oil Change International 2021). In North America, this has led to strong calls under the ecosocialist framework for states to respect Indigenous land and treaty rights (Black et al. 2014) as part of putting an end to fossil fuel development. We saw this in the chapter on social democracy, but here these concerns take on an additional character: they are the most recent instances not of an otherwise amenable capitalist system gone wrong, but of a cruel and rapacious system that has always worked in this way.

Solidarities

A major part of ecosocialist climate response involves the recognition of traditional livelihoods and the principles they offer for more sustainable patterns of life. For example, Shiva (2008) argues that climate solutions should be modelled on already existing low-carbon life-modes of local communities like those characterized by agroecology, traditional knowledge, and food sovereignty. One of the most prominent expressions of these ideas on the global stage occurred during the World People's Conference on Climate Change and the Rights of Mother Earth (2010b) in Cochabamba. Its core concept of living well (or *el buen vivir*) rejected capitalism and set out a series of alternative principles on which to base a climate response, including "the recuperation and revalorisation of the various forms of knowledge, ancestral technologies and local systems of production, distribution and consumption that promote the maintenance of the regenerative capacity of nature."

Magdoff and Williams (2017, chap. 11) take *buen vivir* as an inspiration for the foundations of an ecosocialist society. For them it refers not just to a life in which basic needs are met but also one in which "our activities and relationships and spiritual, mental, and artistic development are diverse and fulfilling, and we are 'in harmony with nature.' Our gains are measured socially rather than through an aggregation of material goods." They envision a society in which all people can participate fully in collective decision-making, including over the economy.

Sovereignty

Wainwright and Mann (2018) offer an intriguing critique of climate politics. They find that even social democratic efforts to see an economic mobilization in the response to climate change work towards the establishment of a *planetary sovereign*, based on a kind of climate Keynesianism (i.e., state management of the capitalist economy) with the perpetual capacity to declare states of exception during which to impose authority in the name of responding to the crisis. They warn that such planetary sovereignty is incompatible with true democracy, and urge alternatives. They identify anticapitalist and anticolonial Indigenous resistance movements as a potential model to restore reciprocal relationships with the land.

Method of Moments

The pursuit of far-reaching social change requires thinking about all the various modes of thought and practice that have to be altered for that pursuit to be successful. David Harvey (2009, 236–47) identifies six in his "method of moments": technology, nature, the activity of production, the sustenance of daily life, social relations, and mental relations of the world. As he has put it, "If we end up in a relation to nature that is materially obnoxious or physically dangerous to us (famines, ozone holes, toxic pollutants, global warming) ... then something ... 'across all moments' — has to change, be it social relations, mental conceptions, everyday life, legal and political institutions, technologies, or the relation to nature." As we think about potential responses to fully address the climate crisis and the broader ecological crisis it is part of, we can use this method to better conceive our interventions, imagining, for instance, the negative and positive feedbacks between changes in the different moments and the different power structures and norms in each.

THE LAST RESORT (?): ANTI-CIV

Before concluding, let us consider one final framework, briefly, that takes a radical, antisystemic critique to an extreme. Anti-civ is an offshoot of anarchism (and for that reason is sometimes called anarcho-primitivism), extending the latter's condemnations of state and capitalist exploitation, authoritarianism, domination, and violence to industrialism and to civilization itself. This framework sees civilization as necessarily unsustainable and as a threat to the possibilities for life on a planetary scale (Jensen 2006a). Anti-civ therefore envisions a political struggle against civilization that includes both nonviolent and militant resistance aimed at disabling key points of civilization's critical infrastructure (Jensen 2006b; McBay, Keith, and Jensen 2011).

The goal of dismantling industrial civilization appears to have limited appeal, however. Even if an anti-civ movement succeeded in gathering support, it would lead to a genocide unlike any in history because so many billions depend on a global industrial system for sustenance and health (Chomsky 2011, chap. 14). (Furthermore, a major blow to the anti-civ movement occurred when Derrick Jensen, its most prominent and prolific thinker, was criticized for making transphobic remarks [Lilac 2013].)

That said, the more militant tactics included under the framework may hold increased appeal should climate change continue to worsen and governments continue to fail to commit to ambitious climate goals. And our justice lens cannot out of hand dismiss the major question at the heart of this framework — whether civilization is inherently unsustainable — though neither can it answer it. The inherency of unsustainability is beyond our capacity to determine with certainty.

CONCLUSION

Due to the short timeframe left in which to act decisively on climate change, ecosocialists have tended to throw their support behind the Green New Deal while ensuring its more radical ideas are not compromised away, particularly those concerning the decommodification of services essential for leading a decent life. Where Green New Deals can be won and these more radical elements can be included, it sets the stage for ecosocialists to pursue even more radical changes in the future that can advance projects that not only defeat neoliberalism but erode the foundations of capitalism itself.

PART 4

··

TO SHAPE A COMING WORLD

10

THE CLIMATE MOVEMENT

WE HAVE NOW SEEN each of this book's ideological frameworks for understanding and responding to the climate crisis. My hope in exploring them is that readers will be able to better situate their own politics in the climate fight. But what ways are there to actually engage in that fight?

Liberal democracies offer shallow, low-intensity forms of political participation in which people have few means of shaping policy. Their most direct form of participation is to select, every few years, which representatives they authorize to make political decisions for them. This process has repeatedly failed to produce governments that are willing to take on climate change at the scale and speed that the imperative of climate justice demands despite the broad democratic support for taking climate action and the viable solutions being presented from the grassroots. To correct for the massive democratic deficit in liberal democracies, the climate movement in the Global North has had to devise a wealth of tactics and strategies to advance climate justice. At the same time, the movement needs to address underrecognized matters of race and inequity.

DEEPENING DEMOCRACY: THE TACTICS OF THE CLIMATE MOVEMENT

Mass Demonstrations: Marches and Strikes

Let's begin with mass demonstrations. The primary point of mass demonstrations is to communicate widespread concern, disagreement, or outrage about an issue that has moral or political urgency. The strategy involves making the demonstrations unignorable so that (a) they attract media coverage and draw widespread popular attention to the issue in question, and (b) the marchers' demands register with decision-makers, whether to pressure the reluctant or bolster the sympathetic. Climate marches take advantage of strategic timing to maximize pressure and exposure either by occurring before politically important moments or by synchronizing internationally during a day (or more) of action.

But the large numbers of people who take part in mass demonstrations create challenges for communicating demands, particularly on complex issues where calls involve more than putting an end to something (as occurs, for instance, during antiwar marches). Though climate marches may bring together people sharing significant concern about an issue, those people also hold a diversity of demands for how it ought to be addressed (Wahlström, Wennerhag, and Rootes 2013; de Moor et al. 2020). And marches are also highly interpretable. What parts of a demonstration will the media highlight or ignore? Is the message that gets picked up what the organizers intended? Or is it something simpler or more anodyne? After all, political elites have little obligation to register a given march as anything other than what we might call mass reserves of raw volition supporting climate "action" to be fashioned into whatever policies those elites see fit. In the 2010s, this became an increasing concern within the climate movement.

In September 2014, special United Nations talks — Climate Summit 2014 — were held in New York to achieve some progress on climate negotiations before the 2015 Paris summit. The moment was fraught. After several years of failure to produce a clear successor to the Kyoto Protocol, a growing awareness had set in that politics as usual was setting the world on the path to climate destruction, and something big was needed as a corrective. A climate march was planned to coincide with the talks, and all signs pointed to it being a massive one. March organizers decided to embrace the inevitable diversity, and they found the key to doing so in an innovation the movement had adopted prominently a few years prior. In 2009, the climate march in Copenhagen, Denmark, was led by leaders from Indigenous communities on the front lines of climate change. The reasoning was that the people who are being impacted first by climate change should be the first to march and represent their demands. Since then, it has become common for frontline communities to lead marches.

The organizers of the New York march, which came to be called the People's Climate March, went all in with this tactic. "In marches as big as this one will be, the many messages we need to communicate to the world often get lost. This time, we want to make sure the People's Climate March clearly expresses the story of today's climate movement — so we're trying something new, and arranging the contingents of the march in a way that helps us thread our many messages together. Together, let's tell the world a clear, powerful story!" said the organizers in an FAQ (People's Climate March 2014).

Altogether, six contingents were arranged to tell that "clear, powerful story." *Frontlines of Crisis, Forefront of Change* featured people who were experiencing the early impacts of climate change and leading the fight against it — Indigenous groups, migrant justice and housing justice groups, and survivors of Hurricane Sandy, which had recently torn through New York and primarily affected communities of colour. *We Can Build the Future* was composed of labour groups, students, families, youth, women, and elders, and was intended to remind people that "every generation's future is at stake and we can build a better one." *We Have Solutions* brought together several environmental groups to show the variety of possible responses to the climate crisis. *We Know Who Is Responsible* assembled protest groups standing against extreme fossil fuel projects like tar sands extraction and fracking, as well as against war and corporate capitalism. *The Debate Is Over* featured scientists and interfaith groups. *To Change Everything, We Need Everyone* demonstrated the diversity of the group beyond the climate movement, bringing in LGBTQ+ groups, and neighbourhood, city, and national groups (People's Climate March 2014). The July 15, 2015, March for Jobs, Justice and the Climate in Toronto, Canada, and the April 2017 People's Climate March in Washington, DC, adopted this kind of line-up as well.

But even with this narrative approach to organization, there remain concerns about the effectiveness of marches. Do they do much of anything at all (e.g., Scranton 2015, 60–68)? Do the demands presented still remain too open to interpretation (Solon 2014)? There has been some thought given to how marches can, instead of dissipating at the end as they often do, use the numbers and strength gathered for something more actionable (Monbiot 2017b, chap. 9).

In addition to communicating demands, mass demonstrations are also celebrations of the movement, showing its strength and diversity. In their inclusivity, climate marches present opportunities for disparate groups to assert concerns about the issue from their respective standpoints. The presence of participants from outside of the environmental movement showcases and celebrates the results of recent organizing efforts to assemble a "movement of movements" (Klein 2014c) modelled on the diverse global justice movement through what is called "frame-bridging" (della Porta and Parks 2014) or "social movement spillover" (Hadden 2014). Migrant justice groups, for example, might understand that climate change will create migration and displacement pressures in a time of border securitization. Food justice groups see that climate change will endanger food security. Animal rights campaigners participate to merge ethical concern for animal life and welfare with concerns about the impacts of the meat industry on the climate. Climate action thus becomes not simply a demand from an undifferentiated mass focused on a narrow environmental issue, but one from diverse sections of society concerned with intersecting matters of justice, so it becomes a broad-based democratic demand stretching far beyond environmentalist circles. And demonstrations also act as a movement gateway. People new to the movement are exposed to local groups that they can join, get educated about solutions and the better world being demanded by the grassroots, and learn about the political forces standing in the way.

Demonstrations also stand as reminders that the movement's politics are alive in the world, and in a big way. In our day-to-day lives, we may not see much awareness or concern about the issues we care most about. To witness thousands assembled on the day of the march is undoubtedly a cure to the isolation — to the sense of shouting into the void — organizers and activists can often feel. The demonstrations also tap into and deploy human artistic creativity in service of free political expression, involving the creation of signs, elaborate props, performance pieces, music, T-shirts, chants, posters, handbills, calls to action, and so on. This artwork is also politically significant. It is an efficient form of message communication, instantly presenting demands in catchy or striking ways. The imagery endures beyond the action in the pictures and videos that last beyond the moment of the demonstration.

School strikes offer a twist on the above. They are inspired by the actions of Greta Thunberg, who, at fifteen, after her native Sweden's hottest summer, began skipping school in order to protest in front of the country's parliament. The tactic was picked up and adopted around the world. As students skip school to protest and make climate demands, it introduces an old and powerful political-emotional element: shame. When children have to rise up to fight for their own future, it lays bare ativev?

the failure of today's political decision-makers to carry out the basic duty of protecting the youth and their future. What is the point of them being in school and preparing for their future if that future will be one of unpredictable ruin? The school strikes also disrupt children's expected role as passive, obedient, and nonpolitical, showing them awakened as a political force.

The school climate strikes lent momentum to a larger climate strike tactic: mass, global days of action in which the strikes expanded beyond students to include workers (Savard 2019), as has occurred on a number of occasions. The biggest so far were the September 2019 actions, where upwards of seven million people took part. Following stoppage due to COVID-19, global climate strikes returned in 2022.

Fossil Fuel Divestment

The fossil fuel divestment movement arrived onto the scene thanks to a landmark essay by campaigner Bill McKibben (2012) in which he helped to popularize the *carbon budget* approach to understanding the climate crisis, which tells us the total amount of carbon that humans can emit before the earth warms by a certain amount. The approach showed that for an 80 percent chance of holding temperature rise below 2°C, humanity could emit only a fraction of the carbon dioxide fossil fuel companies hold in their reserves. This was, as the title of the piece had it, "global warming's terrifying new math." And what made it so terrifying was that fossil fuel companies had every intention of seeing that fraction burned — and more. They were driven by the imperatives of an economic model that valued those companies for the sellable assets they possessed; they *had* to do it. And not only that, there was nothing in the environmental political scene powerful enough to stop them. Environmental groups had for too long urged people to adopt individual lifestyle change in order to take on the climate crisis. A movement was needed. McKibben's hope was that the figures revealed by the carbon budget would direct moral outrage towards the fossil fuel industry, which should now be seen as "a rogue industry, reckless like no other force on Earth." The history of struggle against South African apartheid offered a model of action for wielding that anger: divestment.

Fossil fuel divestment seeks to pressure and persuade institutions (typically pension funds, universities, religious organizations) to stop investing in fossil fuel companies. It simultaneously rallies two overarching sets of motivations, moral and economic (expressed in complex discursive strategies, as described in Mangat, Dalby, and Paterson 2018).

First, divestment activists argue that it is deeply unethical to benefit from investments in corporations whose business models require destructive changes to the climate and impacts on frontline communities. (As the movement sums it up in its slogan, "If it's wrong to wreck the planet, then it's wrong to profit from that wreckage.") And in addition to their impact on the climate, their extraction and refining practices have impacted the health and well-being of nearby communities, often Indigenous and of colour, whose political marginalization limits their ability to stop destructive projects from being located close to them.

Second, as the argument goes, investments in fossil fuel companies will grow increasingly unprofitable as the world gets more serious in taking climate action. If governments set serious and effective policy for meeting the Paris Agreement targets that they have agreed to, then fossil fuel companies will be overvalued; massive portions of their stock will be unsellable, becoming what are called stranded assets. Smart institutions would start taking their investments out of those companies now.

A divestment win in one institution is not on its own meant to change things. Rather, each victory works in concert with others by changing public perception. The general public largely views fuel companies neutrally, as a legitimate industry providing jobs and selling the energy people need in their daily lives. By investing in the fossil fuel industry, individuals and institutions give tacit approval to its activities and preserve its *social licence*, the consent society grants the industry to continue business as usual. But in the same way that a mass sell-off of stocks signals that a company has become financially toxic, so does mass divestment signal that an industry is morally toxic.

The divestment movement started out small. Just 181 institutions representing $50 billion in assets had divested by September 2014 (Arabella Advisors 2015). By December 2021, more than 1,500 institutions, together commanding nearly US$40 trillion, had divested either fully or partially from fossil fuel companies.

A variation on this effort involves using a variety of tactics, from protests to shareholder resolutions, to pressure the financial sector (major banks, insurers, and asset managers) to stop providing capital, loans, and insurance for fossil fuel projects. Just before the emergence of

COVID-19, the "Stop the Money Pipeline" campaign launched with this aim in mind. In North America, targets have so far included Wells Fargo, Blackrock, JP Morgan Chase, and banks and insurers for the Line 3 tar sands pipeline. May 2022 saw the launch of the "Toxic Bonds" campaign aimed at disrupting fossil fuel companies' ability to finance expansion through the bond market.

Occupations and Blockades

This next set of actions increases in directness. Occupations are direct, nonviolent actions where activists hold a space for a sustained period and disrupt its normal functioning in order to pressure decision-makers into taking action on an issue. The spaces to be disrupted therefore need to be chosen strategically. They tend to be sites that are symbolic of the problem (e.g., the racially segregated lunch counters occupied by Black civil rights activists in the early1960s or the New York financial district occupied in the Occupy Wall Street movement in 2011), used by decision-makers (e.g., a politician's or corporate office), or public spaces that, if occupied, grind important functions to a halt (e.g., a major city intersection).

Occupations, if sustained and sufficiently disruptive, require a response. Authorities and decision-makers can accede to occupiers' demands, try to wait out the occupiers, or, if neither of those things happen, call for the state to use force to intervene. This is where the tactic of occupation turns the possibility of occupiers' arrests to its advantage (Monbiot 2019). Arrests make for irresistible news fodder, and the resulting media coverage helps ensure that the occupations, and by extension the cause behind them, reach the broader public. The arrests also convey a moral seriousness; the cause is important enough that occupiers have prioritized it over their own freedom. As the state deploys force, it can slip up and overstep. (In June 2019, on the hottest day in France's history, people were outraged as videos circulated of nonviolent activists being tear-gassed by police. In the United Kingdom in 2019, the London Metropolitan Police issued a blanket ban on Extinction Rebellion protests, drawing wider concerns about civil liberties.) And once it deploys force, the government is put into the position of seeming to defend the immoral action the occupiers are seeking to stop. The priorities of the government are thereby revealed to be backwards, as it chooses to deploy its monopoly on legitimate violence to protect industry, economic growth, or capitalism over the people and their demands. In failing to have attended to the occupiers' demand in the first place, in forcing people to have to be arrested, and in overreacting, the government suffers blows to its legitimacy.

One of the earliest major climate occupations occurred when activists surrounded the White House in opposition to the Keystone XL pipeline intended to transport tar sands oil from Alberta, Canada, to the United States. With the legislative branch of government in the hands of Republicans during the Obama presidency, the movement realized that it would have to put pressure on the executive branch to use its powers to veto the pipeline. Anti-pipeline protests and civil disobedience actions at the White House — the first in September 2011 when ten thousand protestors surrounded the building, another in March 2014 — sought to challenge the integrity of President Obama, who had campaigned on climate issues.

Another notable occupation occurred in November 2018 when activists with the Sunrise Movement and the Justice Democrats staged a sit-in at the congressional office of Nancy Pelosi one week after the Democrats retook the House following the midterm elections. Mainstream Democrats like Pelosi, who was vying for House majority leader, had for too long been unwilling to take serious action on climate. The activists were demanding the creation of a Select Committee on a Green New Deal that would have power to draft legislation. The fifty-one arrests, and the appearance by newly elected congresswoman Alexandria Ocasio-Cortez in solidarity, massively raised the media profile of the action, and with it the Green New Deal.

The 2018–19 occupations organized by Extinction Rebellion brought the occupation tactic to its highest intensity yet, particularly in the United Kingdom, where arrests numbered in the thousands as activists blocked major bridges, road junctions, and public squares. October 2019 alone saw 1,400 arrests. Occupations continued right up until the social distancing and lockdown measures began in response to COVID-19. The most prominent was probably the "Stop the Money Pipeline" actions that took place on Wall Street in early 2020 to draw attention to the role that major private banks play in financing fossil fuel companies.

Blockades are direct nonviolent actions using bodies, built structures, or both to deny the start or continuation of activities deemed immoral or destructive. Unlike with an occupation, participants do not aim to win widespread popular opinion. Blockades are not set in symbolic spaces, but in locations where they can stop an activity from occurring, like construction or extraction sites (or routes to them). They arise from an urgency to protect a community or ecosystem from an impending threat. Wider popular opinion might eventually come to support the demands animating the blockade, but that is of secondary concern. However, blockades are often supported more widely by political allies, who can take solidarity actions.

Naomi Klein has popularized the term "Blockadia" (originally coined by activist group Tar Sands Blockade). As she put it (2014a, 294–95), "Blockadia is not a specific location on a map but rather a roving transnational conflict zone that is cropping up with increasing frequency and intensity wherever extractive projects are attempting to dig and drill, whether for open-pit mines, or gas fracking, or tar sands oil pipelines." In North America, so many of the blockade actions have been led by Indigenous communities. There is in these instances a powerful intersection between ongoing colonial control, the fight for survival that Indigenous people have been waging for half a millennium, and fossil fuel extractive infrastructure.

The blockade that probably holds the highest place in the history of the climate movement was sparked in 2016 by the resistance led by the Standing Rock Sioux tribe against the Dakota Access Pipeline, intended to transport crude oil from North Dakota's Bakken oil fields to refineries in the state of Illinois. If it leaks, the pipeline will issue oil directly into the Missouri River, the tribe's main source of drinking water. In September 2016, construction crews began bulldozing lands the tribe considers sacred. Water protectors attempted to block further work and were set upon by dogs and pepper-sprayed by security personnel. As media showed video of the attack, the #NoDAPL movement grew in force and brought thousands of people — Indigenous people from across North America, environmentalists, even US army veterans — to the blockade encampments, where they would withstand repeated incursions from law enforcement through the months that followed. (Though Obama would order the pipeline construction to be halted, Trump rescinded the order in early 2017 and state authorities cleared the encampment in February, and oil started flowing by summer of that year). In eastern Canada, Elsipogtog First Nation endured Royal Canadian Mounted

Police (RCMP) attacks on their peaceful blockade against fracking projects on unceded territory. In British Columbia, the Wet'suwet'en Hereditary Chiefs have used blockades to prevent the construction of the Coastal GasLink natural gas pipeline through their unceded traditional territories. In the same province, the Tiny House Warriors built small homes on unceded Secwepemc lands along the route of the Trans Mountain expansion pipeline intended to carry Alberta tar sands oil to the Pacific coast.

Legal Challenges

The number of climate litigation cases around the world has been growing quickly (United Nations Environment Programme and Sabin Center for Climate Change Law 2020), and the movement has been involved in a number of different types of them. The first and probably most prominent type so far involves lawsuits aimed at forcing governments to take on more ambitious climate targets. One variant involves arguing that governments are violating or failing to protect human rights unless they take on ambitious climate policies. In the Netherlands, the citizens group Urgenda (short for "Urgent Agenda") brought a case to the courts arguing that, by not reducing emissions enough, the Dutch government was in violation of Articles 2 and 8 of the European Convention on Human Rights, which require the protection of the right to life and the right to respect for private and family life, home, and correspondence. After a years-long legal battle, the Supreme Court in December 2019 upheld a lower court ruling that the country had to reduce emissions 25 percent compared to 1990 levels by 2020. The ruling had the government scrambling in 2020 to issue a series of policies that could bring emissions down including through restrictions on the capacity that coal plants could operate at. In April 2021, Germany's supreme constitutional court ruled in favour of a group of young environmental activists and nongovernmental organizations and decreed that the government had until the end of 2022 to strengthen its climate plan, its existing one being too weak to protect the freedom and rights of young people. A second variant of this first type of legal action attempts to force governments to comply with preexisting climate legislation. In 2020, the Supreme Court of Ireland ruled in favour of advocacy group Friends of the Irish Environment, ordering the government to quash its insufficient 2017 climate plan and pass another. (The ruling was due to the plan falling well short of the country's 2015 climate change act requiring 80 percent emission reductions by 2050.)

A second type of legal challenge involves groups using the courts to fight fossil fuel development projects from proceeding, due to the risks they pose or damage they have already done. In the United States, the Standing Rock Sioux turned to the courts to stop the Dakota Access Pipeline; in 2020, a judge ordered a new environmental impact assessment of the pipeline, which could require it to halt operations. In Alberta, Beaver Lake Cree Nation is taking on the provincial and federal governments. The nation is arguing that permitting tar sands extraction on its traditional territories has severely impacted the local environment on which the community depends for food and water, putting the governments in violation of treaty law. The case is expected to go to trial in January 2024.

The third set of legal actions has involved citizens attempting an innovative strategy against criminal charges resulting from taking direct actions: the *necessity defence* (Climate Disobedience Center n.d.). To put it simply, the defence is based on the argument that it is now a necessity for ordinary people to take extra-parliamentary and extra-legal measures to advance the effort to drastically reduce emissions. This is because the climate crisis forms a massive and unique threat and because governments have failed to respond to it (and are even approving new fossil fuel projects). Laws that protect fossil fuel property are effectively protecting the processes destroying conditions for human well-being. We might consider this set of actions as attempting a kind of fix for a flaw in the system. In the same way that carbon pricing corrects for a negative externality that underprices fossil fuels, direct actions taken under the necessity defence are a corrective for a legal system that is overprotecting projects destroying human rights. The necessity defence has been attempted by defendants in a number of cases, but with mixed results. In 2008, Greenpeace activists in the United Kingdom who defaced a smokestack were acquitted on a version of the necessity defence (Vidal 2008). In the United States, activists who coordinated across four states (Washington, Montana, North Dakota, and Minnesota) to turn manual shutoff valves on tar sands oil pipelines all planned to use the defence. Only the Minnesota activists were permitted to use the necessity defence, though it went untested when the trial was thrown out due to prosecutors not being able to prove the activists actually caused any damage (McGraw 2018). In Canada, activists in British Columbia went to trial for defying an injunction making it illegal to enter an area where they would have perturbed construction on the Trans Mountain expansion tar sands pipeline. The presiding judge refused to allow the necessity defence, for rather unconvincing reasons (Saad, 2019b). In April 2021, during a trial of protestors who broke windows and graffitied Shell's headquarters in London, a UK judge told jurors that the protestors had no defence in the law. They were nevertheless acquitted.

Party Politics and Elections

Beyond these means, activists have also been organizing to more directly influence legislative proceedings. Following the defeat of Bernie Sanders in the 2016 Democratic primaries, progressive political action groups Brand New Congress and Justice Democrats looked for ways to reshape the staid Democratic Party still stuck in the mire of neoliberal orthodoxy. They recruited Alexandria Ocasio-Cortez and canvassed for her in her successful campaign to unseat an unchallenged centrist Democrat. (One of her earliest acts was to join the Sunrise Movement in their occupation of Pelosi's office described above.) The rise in prominence of Green New Deal–type programs has made a difference in these efforts. These programs are effectively justice-based political platforms that emerge from the grassroots. Electoral candidates simply have to agree to embrace them. Some of the organizing around Green New Deals in recent years have seen climate justice groups endorse candidates. For instance, in the 2019 and 2021 Canadian federal elections, the Canadian arm of 350.org presented a list of candidates it endorsed for embracing ambitious and just climate action.

"DAMN NEAR BLEEDING TOGETHER": THE MOVEMENT AND RACE

In 2020, 23-year-old Ugandan climate activist and founder of the Rise Up Movement Vanessa Nakate was cropped from an Associated Press photo that, when published, featured only young white climate leaders at a major press conference. On a video posted to Twitter, Nakate (2020) described it as an instance of how representatives of views from

Africa are devalued, and as "the first time in my life that I understood the definition of the word 'racism.'" The Associated Press stated that the cropping was done for compositional reasons (behind Nakate is a wall, while behind the others is an unobstructed view of the mountains). The incident was an ugly metaphor for whose voices get to be presented to the world and for how some concerns do not fit into a particular conception of what the climate fight should look like. In recent years, increasing attention has been paid to issues of race within the climate movement because moments like these have become too common.

For instance, in a London climate march just before COP21 in 2015, Wretched of the Earth, a bloc representing communities of colour on the front lines of climate change, had been promised the lead spot. But in the end, it seemed as though their messaging, which focused on how colonialism and corporate capitalism engendered the climate crisis and created differential impacts on communities of colour, was too controversial and divisive for march organizers. Wanting to present a friendlier face, organizers arranged instead for activists in animal costumes to make their way to the front (T. Brown 2015; Kelbert and Virasami 2015; Wretched of the Earth 2015).

In February 2020, a piece in *Vice* caught some attention under the title "Why I Quit Being a Climate Activist: The Climate Movement Is Overwhelmingly White. So I Walked Away" (Hermes 2020.) A sour reaction followed online, its general sentiment captured in one Reddit user's sarcastic comment, "Yes, I turned my back on the notion of saving the planet because too many white people were doing it. Great logic."

But much of that reaction seems to have missed the point. The author, a Berlin-based climate activist of Filipino-German ancestry, makes clear it was not the overwhelming presence of white people per se that pushed her away, but the psychologically and emotionally exhausting dynamics that a person of colour too frequently experiences in predominantly white spaces. They include repeated failures to meaningfully address concerns about including questions of race in the larger struggle; persistent negation or unawareness of the links between race, capitalism, and climate change; and pervasive tokenistic inclusion of people of colour at climate events.

That tokenism can arise from failures, however unintentional, to shape the movement so that the concerns of people of colour rank among its top priorities. Not appreciating how these failures can reduce the movement's appeal, well-meaning activists attempt to diversify it on the basis of implicit assumptions that people of colour have simply not been recruited. Climate justice activist Payal Parekh (2019) shared a telling message she received through Facebook from Extinction Rebellion: "I'm contacting you on behalf of #ExtinctionRebellion and was wondering if you would be interested in officially supporting the movement? We're looking for support to increase our visibility and reach, and we cruelly need people that are not white men." Parekh responded on Twitter,

> Sorry @ExtinctionR, thx for asking me whether you can use me to show that your "rebellion" is diverse, but no thanks. Maybe it is time to think about why there are few PoC [people of colour] in the North taking part and why there are so few groups in the South. I would posit that had you built up #ER [Extinction Rebellion] with a diverse group from the beginning, then how you function & the actions you do would look different and would have a much wider reach.

These dynamics can impose a sense of isolation and alienation. Mary Annaïse Heglar, cocreator and cohost of *Hot Take*, an American podcast using a "feminist, race-forward lens" to talk about climate change, said the following in a 2020 episode:

> I personally feel like I find myself in this position where I am kind of looked to in the climate movement as the climate movement's 'Black friend.'... That is not the service that I'm here to perform.... I've seen a lot of people circulating my Green Voices of Color list.... It sort of seems like people think I created that list so that white people could then find people of colour.... I created that list so people of colour could find each other, actually. I created that list because I felt lonely. And I felt like a lot of white folks were basically treating me like their 'Black climate friend.'

> After speaking on a panel — and I know a lot of other women of colour have had this experience because I've heard them talk about it — you speak on a panel as a woman of colour and all of these white women are waiting for you as soon as you get off the stage and they're like in tears about how much you moved them

and how much you've taught them.... Now you've got to console them.... And on the other side of her is all these other women of colour that you actually want to connect with. (Heglar and Westervelt 2020)

To be sure, there have been moments when prominent voices in the climate movement have asserted the importance of foregrounding struggles for racial justice (Klein 2014d), including Extinction Rebellion cofounder Stuart Basden (2019). But a general underappreciation of matters of race within the movement has occasionally led to some problems (Gayle 2019; Wretched of the Earth 2019; Táíwò 2020). We can identify at least three things to watch for.

First, strategies for building the movement and increasing its impact can be based on failures to understand how carceral state power reacts and discriminates based on race and Indigeneity. A major part of Extinction Rebellion's strategy, for instance, centred on mass arrests. Instead of recognizing how this might put members of communities targeted by a racist legal system disproportionately at risk, members of Extinction Rebellion seemed more concerned with finding ways of engaging amicably with police. These included sending flowers and thank-you notes to police stations (Blowe 2019), snitching on immigrants suspected of carrying out thefts during demonstrations, or suggesting to police that their services are better deployed in high-crime areas (which happened to be racialized communities).

Second, language and messaging about the climate crisis can elide important historical and contemporary injustices. Framing the threat of climate change so that it overemphasizes coming social breakdown can gloss over the ways that communities of colour have already faced and are currently facing social breakdown in the forms of past and present colonial rule, institutional racism (through being underserved and over-policed, facing racist voter laws, etc.), and indeed the already occurring and disproportionately felt effects of climate change. It raises questions of whose social breakdown matters — and matters enough to take to the streets to prevent. Another issue concerns messaging around climate change–impelled migration. There is a major difference between framing people migrating and being displaced due to climate change as parties being owed rights, protections, and redress on the one hand (Saad 2017) and as threats to the societies of the Global North on the other.

Third, the movement's analysis of the climate crisis can ignore histories of racism, colonialism, and unequal development. Movement members from more privileged communities can be frustratingly incurious about how these histories are implicated in having made fossil fuels accessible for exploitation and in creating vulnerabilities to climate change in communities of colour around the world.

Some priorities that can foster a more diverse and inclusive justice-based climate movement have been proposed (Rahman 2019; Wretched of the Earth 2019; Guardian 2020). For one, the movement should more prominently show recognition and concern about the impacts of climate change on — and raise the voices of — communities in the Global South who are dealing with the legacies of underdevelopment and communities in the Global North, so often of colour, who are dealing with settler colonialism, systemic racism, and marginalization. Second, movement analysis and actions need to identify and target the major corporations that are not just driving the climate crisis but are responsible for violence against Indigenous land defenders and environmentally destructive resource projects throughout the Global South. Third, and relatedly, the movement should highlight the historical and ongoing role of colonialism in environmental degradation. Finally, the movement should act in solidarity with people of colour and support groups and actions fighting for migrant rights, decolonization, and antiracism.

The 2020s began with a confluence of events that gave the climate movement an opportunity to look with new urgency at issues of race. In 2020, the brutal killings of George Floyd by Minneapolis police and of Breonna Taylor by Louisville police set off a renewed wave of Black Lives Matter protests throughout the Western world that pushed questions about the relationship between police and communities of colour to the top of the political agenda. At the same time, COVID-19 spread throughout the world. In the United States, the federal election heightened political tensions. Essays and think pieces abounded using lenses more sensitive to the links between these matters (e.g., Brecher 2020a; Farand 2020; Perry 2020).

Systemic racism was the common thread running through the violent, unaccountable overpolicing of which Floyd and Taylor were victims, the poverty and underservicing that was making the pandemic more pronounced in communities of colour, the historical siting of toxic pollution near communities of colour, and the long-standing underinvestment that made those communities more vulnerable to early effects of climate change. The democratic impulse to shape public spending that animated the Green New Deal was the same one behind calls to shift funding away from police and towards alternative community priorities. And there were signs of trying to do better in parts of the climate movement that had underappreciated the oppressive relationship between police and people of colour when using the strategy of mass arrests. Extinction Rebellion (2020b) issued an apology and promise to bring issues of race to the forefront. Heglar perhaps summed up the moment best on *Hot Take*: "You don't have to play this game of 'my disaster's bigger than yours.' ... You can just connect the dots because they're not unconnected; they're damn near bleeding together" (Heglar and Westervelt 2020).

BEYOND PACIFISM

In 2021, historian Andreas Malm published a deliberately provocative book called *How to Blow Up a Pipeline* that could create a turning point in the climate movement. The project emerged from Malm's (2021b) observation that the climate movement has by and large remained wedded to an unevolving set of tactics like those described above, even as it became larger and more powerful than ever before (the school strikes and the rise of Extinction Rebellion being two significant examples). But surely, given the urgent existential threat of the crisis — and the ongoing failure to win policy recognizing it in these terms — the movement ought to be willing to experiment with more direct methods.

To successfully respond to the climate crisis, much of today's energy structure will need to be decommissioned long before its various owners and investors realize their profit (or even break even). But the states of the world are unwilling to ever pursue this. Property holds "the status of the ultimate sacred realm," and as such, he states chillingly, "property will cost us the earth" (Malm 2021a, chap. 2). Malm argues that it therefore becomes the task of ordinary people to declare and enforce a prohibition on new fossil fuel infrastructure, to damage it or otherwise put it out of commission where it appears; it is this always looming threat to profits that will be required to turn otherwise recalcitrant capitalists away from investing in the property and processes further driving the climate crisis (Malm 2021a, chap. 2; 2021b).

He finds that the history of militant struggle in the two hundred years from the time of the French Revolution to the fall of the Berlin Wall has been not simply forgotten but actively denied, leading to a deskilling of movements with respect to militant potential. Malm seeks to unsettle the dominant narrative told by mainstream voices — in academia and in activist circles — insisting that nonviolent tactics hold a superior historical record of winning positive social change compared to militant ones. It's a narrative that leans heavily on several standard examples of bold nonviolence being the pathway to victory: the abolitionists bringing slavery to an end, women suffragettes in the United States winning the right to vote, Gandhi's nonviolent *satyagraha* movement for Indian liberation from British imperial control, and the American civil rights movement led by Dr. Martin Luther King Jr. These examples have come to inspire the climate movement, with Gandhi standing as a model for many climate activists (Malm cites Bill McKibben calling Gandhi "our scientist of the human spirit, our engineer of political courage"). But, Malm notes, there are problems in these (and other) cases. For instance, Indian liberation required a world war to drain the British empire's imperial resources, and Gandhi's movement was hardly the only form of resistance against the colonizers. In the case of nonviolent Black civil rights activists, they were protected by armed Black members to scare off any would-be attackers.

McBay, Keith, and Jensen (2011) were probably the last environmental writers to gain some prominence in taking up the question of militancy versus nonviolence. Associated sometimes with the deep ecology movement, they advocated a militant vanguard movement that had as its goal the collapse of civilization itself. Malm has been careful to distance his proposals from that anti-civ path, insisting they are part of a struggle *for* civilization, one not based on fossil capitalism.

The pragmatics of climate militancy — whether it would be feasible for activists to actually destroy fossil fuel infrastructure — are not the issue. Malm notes multiple cases of resistance movements targeting precisely that infrastructure, not for reasons of climate change, but nonetheless effective at taking it out of commission. At question, rather, is morality and political effectiveness.

Let's begin with morality. To be clear, Malm is not endorsing a blanket use of violence in the name of saving the climate. He insists that an escalation of tactics should never reach the point where they target human life. (As he puts it, "attacking the physical property, the machines, the infrastructure, the installations, that destroy our planet, is something very different, qualitatively different, from attacking the body of a human being. So, I hope there's no doubt about me totally rejecting methods of assassination and murder" [Malm 2021b]). It is property alone that would be the target, and the words *vandalism* and *sabotage* capture the sense of the activity he has in mind rather than something like *terrorism*. With those restrictions in mind, is there any moral reason to value fossil fuel property over the climate given that they stand in opposition? One would be hard-pressed to offer anything convincing.

The matter of political effectiveness is more difficult to address, however. What effect would a more militant part of the climate movement *really* have? Malm refers to the "radical flank effect," in which nonviolent tactics only work because — in the estimation of the state — the part of the movement that nonviolent activists represent becomes the "reasonable" alternative, the lesser evil that can be tolerated and negotiated with compared to the radicals who will leave little to nothing of the status quo. (In this, he continues a tradition probably best known through Ward Churchill's [2007] *Pacifism as Pathology*.) Malm also suggests that sabotage of fossil fuel infrastructure will be more impactful and acceptable to the wider public if it is timed to coincide with devastating climate disasters.

But he is also open to the possibility of a *negative* radical flank effect. This would occur if increased militancy makes the climate movement appear so distastefully extremist that it loses political influence. In the worst case, it would alienate a host of people precisely at a moment when the climate movement cannot afford to become less effective. This could potentially be addressed as long as that radical flank adheres to careful principles and is prepared to call off further property destruction should it "draw too much retaliation, vilification, embarrassment on the movement" (Malm 2021a, chap. 2).

A more militant climate movement could also experience a negative flank effect if its activities give state authorities impetus to crack down on any and all protest activities. It would, in this worst-case scenario, be a gift to the forces of repression looking for an excuse to beat back, arrest, and punish uppity activists. (And one can also easily imagine how a far-right media ecosystem would whip up support for this repression upon finding out a migrant or person of colour was among militants destroying property.) Indeed, those forces of repression already exist, and one critique of Malm's work is that it ignores the harsh consequences from the state awaiting those who would follow his program (Wilt 2021). For instance, internal documents obtained from the RCMP showed Canada's federal-level police force characterizing climate activists — whom the report alarmingly referred to as "violent anti-petroleum extremists" following an "anti-petroleum ideology" — as posing domestic threats to the oil and gas industry (Linnitt 2015). In 2019, when the RCMP strategized how to clear the blockade established by Wet'suwet'en land defenders in British Columbia to prevent the construction of a natural gas pipeline through their traditional lands, commanders argued for deploying personnel prepared to use lethal force (Dhillon and Parrish 2019). In 2020, the *Guardian* broke the story that the UK terrorism police had included Extinction Rebellion as an extremist ideology (alongside neo-Nazi terrorism) that government organizations, police, and teachers are obliged to report to the United Kingdom's counterterrorism program Prevent (Dodd and Grierson 2020). In the United States, the late 2010s and early 2020s saw states propose or pass a series of antiprotest laws, including laws seeking to penalize protests against "critical infrastructure" like fossil fuel pipelines (Mueller-Hsia 2021), and some that would allow people to, without penalty, drive their cars through protest groups blocking roads. In 2016, climate activist Jessica Reznicek sabotaged equipment and material for the Dakota Access Pipeline. After she entered a guilty plea, federal prosecutors convinced a judge to increase her sentence using a "terrorism enhancement" measure (Bruggeman, Dwyer, and Ebbs 2022). In spring of 2021, the United Nations special rapporteur on the rights to freedom of peaceful assembly and association announced he was conducting an investigation into state and private repression of the climate justice movement.

Let's close on a thought experiment Malm proposes. We can imagine a first scenario in which the nonviolence of today's climate movement does in fact succeed in reversing the crisis. All well and good for the pacifists. But what about a second scenario in which, a few years from now, the generation that learned of the climate crisis through school strikes and the like look around and find that that crisis is still ongoing? Emissions, here, are still on a pathway well out of line with what is needed to preserve a safe climate. "What do we do then?" Malm (2021a, chap. 1) asks. "Do we say that we've done what we could, tried the means at our disposal and failed? Do we conclude that the only thing left is learning to die — a position already propounded by some — and slide down the side of the crater into three, four, eight degrees of warming? Or is there another phase, beyond peaceful protest?"

CONCLUSION

It is the nature of democracy to resist constraint by insufficiently partic-ipatory decision-making institutions. In attempting to compensate for the democratic deficit — to build the ramp leading up and across from a fossil-fuelled today to a postcarbon tomorrow — the climate movement has innovated and assembled within its toolkit a wide array of tactics. A set of related questions follows. How do the tactics surveyed in this chapter fit with the ideologies we have looked at in this book? Are some in service of preserving the systems of the status quo (minus the fossil fuels) while others are pushing towards more utopian visions of system change? Would there be a relatively clear split between which actions moderate and radical activists choose to engage in? Should we expect that system-fixing centrists are attracted to actions operating on the be-lief that the existing order does not need to change and that it will re-spond to the crisis given more popular pressure — actions like marches, divestments, legal actions, and the occasional occupation? On the other hand, will it primarily be system-replacing leftists gravitating towards tactics that seem to operate on the belief that the existing neoliberal or-der will never act against the wishes of its elite ruling classes — the direct actions of blockades and, perhaps in the offing, militant sabotage?

The literature does not yet provide a clear picture. Even without con-sidering the more militant actions just discussed, scholars have argued that climate movement tactics have the potential to uphold or erode the existing system (Stuart, Gunderson, and Petersen 2020). My own expe-riences in the climate movement suggest that those holding more leftist beliefs will seek to explore the most radical potentials of each tactic and use those tactics to amplify system-critical messaging rather than hold up their noses at less direct action. There is too little time left to avoid committing the earth to more than 1.5°C of warming to seek only to build power for radical actions.

Even if many of the tactics reviewed here can be employed without challenging the system, there are also ways to use them to shame the individuals and discredit the ideologies upholding and reproducing the dynamics failing to address the crisis. Critical questions about the nature of the system that activists are engaging with are not far below the surface of the strategies they take part in. After joining in marches and climate strikes, activists might begin to wonder about the deeper nature of the reigning (and ostensibly democratic) institutions that continue to ignore them and the crisis. Divestment activists in North America need not look too deeply to observe how the fossil fuel production, pipeline, and refinery companies they seek to pull investments from are closely allied with the state to jointly perpetuate the colonial practice of prioritizing capitalist resource extraction over Indigenous land and life.

In closing, readers eager to get involved in the climate struggle have a wealth of options compatible with a range of ideological beliefs.

CONCLUSION

....

WORLDS AT STAKE

IN TWO MASSIVE WAYS, the climate crisis places the world at stake.

The first is the more obvious. Every blistering heat record, every astonishing new sea-ice minimum, every unprecedentedly hellish wildfire, every aberrantly wrathful storm is a testament to the collapse of the climate that has been a defining feature of our world since the end of the last ice age. In this *environmental* sense, our world is very much in trouble.

The other sense in which the climate crisis puts worlds at stake is *social*. Deciding what is to be done about the climate emergency involves major choices about the nature of the economy, the state, social inequality, individual freedom, rights, sustainability, and more. These choices open up very different possibilities for how life is to be lived. What social world will be ushered in, protected, or prevented from coming into being as we take on the emergency?

It's these different worlds that might be made real or denied — these worlds at stake —that we looked at in this book. The introduction noted that climate change signals an alarm that something about our way of life must change, and people hear this alert differently because they hold contrasting systems of political beliefs concerning the nature of a good society. Chapters 4 through 9 focused on how different ideologies hear that climate alarm. The closing sections of those chapters raised what I consider to be important critiques that emerge from a climate justice lens.

For neoliberals, who are theoretically opposed to government intervention in the economy, finding policies to address the climate crisis long posed a problem. They have tended to come to terms with it by shaping it into a problem for the market to solve. Climate change becomes the result of a pervasive market externality: the unpriced cost of carbon emissions and the consequent underpricing of fossil fuels. Once that externality is internalized, the market can provide the right signals to individuals, governments, and corporations, all of which will now be incentivized to shift production and consumption away from carbon-intensive fuels, products, infrastructure, and systems in favour of greener alternatives. Beyond these steps, there is little need to tinker with the existing neoliberal order still dominant throughout the world, contrary to what more progressive or radical voices might urge.

Using our climate justice lens, we drew attention to how this neoliberal response might be system protecting but climate sacrificing. Perhaps if the suite of market-oriented solutions — chiefly carbon taxes, emissions trading, and light regulations — had been fully and properly deployed in a previous era of climate politics, we would not be facing a crisis today. Perhaps, having begun at the turn of the millennium, this gradualism might have sufficed to reconcile capitalism's need for economic growth and profitability with the need to achieve economy-wide clean energy adoption, all while leaving the heaving treasure mounds of corporations and the richest 1 percent untouched with plenty of opportunity to grow higher still. Perhaps. But as the years of the 2020s tick away, attempting to reconcile economic growth, rapid clean energy adoption, low wealth taxation, and a sharply unequal neoliberal class order — all of it in a single package of nondisruptive social change — appears to have proven too difficult. The neoliberal framework thus leaves us in doubt about the degree to which it will prioritize ambitious climate action.

For some, even that market-based neoliberal response goes too far. Chapter 5 focused on the most influential right-wing response over the past generation: climate change denial. As late into the climate crisis as the 2020s, denialism continued to have enormous political influence, most significantly through the US Republican Party. For deniers, the world at stake is one in which a particularly rigid vision of society compatible with right-wing ideology remains coherent. If the science illuminating how humans are driving ever more dangerous climate change is permitted to be believable, then a vision of an ideal society in which governments refrain from intervening in our lives becomes increasingly implausible.

If the struggle for the climate is to be won, the denialist project is one that must be driven to the fringes and there immured, unable to re-emerge as a political force. The main players in the denial machine — the funders in the fossil fuel industry and conservative foundations, the producers of misinformation in the right-wing think tanks, and the influencers spreading denial in the corporate and alternate right-wing media — have together carried out one of history's most morally grotesque endeavours. And it has not stopped, even as the impacts of climate change are obvious and widespread. Alongside the "classical" forms of denial, recent years have seen growing emphasis on solutions denial, which urges delay in ways that are subtle but also deeply insidious, spreading doubt about the actions that can be taken on climate change during the last moments when transformative, highly ambitious policies could keep globally averaged warming well below 2°C. Denialism, no matter the form it takes, deserves no seat at society's negotiating table. So what should be done about it? How best to expel it from mainstream political conversation is the problem to be solved. But what response can navigate issues of well-funded political propaganda exploiting the rights to free speech across for-profit corporate news and social media? Chapter 5 considered possibilities for undermining the political economy of outlets like Fox News, but would that be enough? And what is to be done about social media? At the same time, the spectre of ecofascism looms, offering radicalizing portions of the right wing a means by which to accept the reality of climate change because of how responding to it can advance an increasingly supremacist ideology.

Chapter 6 explored the geoengineering framework. Its embrace of technologies that would directly and intentionally alter the workings of the climate system itself raises no shortage of concerns through a justice lens. To what degree can such decisions be democratic, in the fullest, most participatory sense of the word — particularly if we will be pressed to make them out of a sense of desperation caused by the failures to overcome neoliberal incrementalism and right-wing denialism (both, in their own way, advancing the interests of the fossil fuel industry)? To what degree would such technologies be the result of a dangerous combination of hubris and ideological recalcitrance?

We referred to these three frameworks as system preserving. For all their differences, each not only offered its own ways to protect large parts of the status quo and existing arrangements of wealth, power, and privilege but also complemented each other in doing so. Denialism could pound the advance of any proposed climate solutions with the flak of propaganda and rally opposition from a significant portion of the right-wing population. Neoliberalism, meanwhile, could hold the line against the transformative — limiting responses to individualist steps and, eventually, incrementalist market-oriented policies that have never quite recognized the severity and urgency of the crisis. It also benefited from the agitations of deniers to the right, in contrast to whom neoliberals at least had solutions. And now, in this late hour, geoengineering offers the tantalizing hopes of a last-minute technological saving grace from our future-bringer, investor-innovator-entrepreneurial class who claim to see the pathway to the coming age differently and more clearly than us mortals.

(Something we did not explore in that section of the book is the possibility of a system-preserving ideological framework that advocates for the state to drive an economic mobilization to address emissions within a capitalist economic system, but has little concern for promoting justice or equality or even sustainability beyond climate change. Could coming years see something of that sort? While such a project is not yet clearly in play, perhaps it could come about through a powerful capitalist bloc confident that its profitability would not be harmed — or, better, certain it would be enhanced — by an aggressive transition to a postcarbon world. This bloc might, at the same time, be deeply concerned about the threats to its profitability coming from worsening climate impacts. Key segments of this bloc would be major players in green and nuclear energy, geoengineering, and agriculture industries. The bloc would make lavish contributions to political parties that promised them lucrative government contracts, subsidies, and beneficial regional or global trade agreements — and who were not overly concerned with funding social programs or enhancing labour rights or regulations. How such a party might appeal to an electorate, what values it would forward through its branding and campaigning, would need some thought.)

Because the neoliberal and geoengineering frameworks raise so many questions, it was essential that we looked at other, system-changing frameworks, other worlds that we might hope to bring about instead. In

Chapter 7, we looked at social democracy and a society-wide rapid economic mobilization for a safe climate under a Green New Deal. Though it would work from within capitalism, it holds system-transformative potential through the enlarged role it sets for the state in responding to climate change at the necessary scale and speed while undoing core elements of the neoliberal order dominant over the last generation. The 2020s started with conditions that opened up political space for a Green New Deal — the fresh vision and energy of the Green New Deal itself, a powerful and optimistic climate movement, a pandemic-caused recession that large public investment could ease, and, in the United States, an incoming administration ready to implement some core Green New Deal policies. However, a strong Green New Deal–like program has yet to take root. If one does and sets up a model inspiring other democratic movements abroad, there might still be time for this to become a Green New Decade, potentially putting us on a path to prevent average temperature rise from exceeding 1.5 or 2°C. A couple matters raised through our climate justice lens would need to be addressed though. One is the potential for new colonial impulses to find expression as demand rises for the materials to power a capitalist postcarbon world. The other is that, even if it manages to solve the problem of greenhouse gas emissions, a Green New Deal as typically presented would still leave intact the mass consumption that drives other ecological crises.

That last matter was explored in Chapter 8. At its core, degrowth attempts to imagine a richer life beyond a society of perpetual, compound economic growth. It reconceptualizes the good life so that it can be lived well by all within ecological boundaries. In doing so, it urges us to think about the nature of the economy more intentionally, to ask what is really essential for well-being. With that more deliberate and discerning economic eye, we could demand much bolder action on climate change by eliminating portions of the economy that drive growth and consumption but do little for well-being and require energy and material throughputs that contribute to ecological crises, climate change being a preeminent example. Through our climate justice lens, we raised a few matters for concern. For one, would important social values and goods found in liberal capitalist societies — cosmopolitanism, scientific progress, individualism — be lost in a degrowth society? For another, what exactly is the political path to degrowth? If a postgrowth world is one that would improve the human good, should the road to it not be better theorized?

In Chapter 9, we focused on ecosocialism, which combines the classic left-wing social critique of capitalism with ecological critiques of that system. Given the short time left to respond to the climate crisis, ecosocialists have tended to add their voices to calls for a Green New Deal but stress its most system-transformative aspects, particularly around decommodification of human necessities. The socialist tradition can offer means of refining our climate justice lens by drawing our attention to questions of solidarity, reparations for capitalism's historical injustices, and the wider collection of interacting thoughts and practices (or "moments") that will need to be considered if we are to bring about just and far-reaching political change.

In the last chapter, we surveyed the actions of the climate movement, theorizing how its diverse strategies work to advance progress on climate action. We used our justice lens to focus on the importance of incorporating and foregrounding questions of race, colonialism, and environmental justice in taking climate action and analyzing the drivers and differentiated impacts of the climate crisis. The chapter closed with a discussion of how the movement deals with the question of political moderation versus radicalism in its actions, noting that even indirect actions can hold more radical implications offering activists of all stripes a diversity of ways to get involved.

OVER TO YOU

What world should we bring about as we take on the climate crisis? It is my hope that, through this book, readers have found themselves better able to answer this question and to find their place in this struggle. Because the future world really is at stake in these next few years. It is right now that so much will be determined with respect to how human life can and will be lived on this planet alongside everything we share it with. It is a tremendous responsibility to shape a coming world. And we — those of us here now — are the only ones around to decide what kind of world that will be.

REFERENCES

Albert, Michael. 2003. *Parecon: Life After Capitalism*. London: Verso.

Albert, Michael J. 2020. "Capitalism and Earth System Governance: An Ecological Marxist Approach." *Global Environmental Politics* 20, 2. https://doi.org/10.1162/glep_a_00546.

Alston, Phil. 2019. *Climate Change and Poverty: Report of the Special Rapporteur on Extreme Poverty and Human Rights*. Geneva, Switzerland: United Nations Human Rights Council.

Anderson, Dave. 2019. "Who's Behind Trump's Claim the Green New Deal Will Cost $100 Trillion?" *Energy and Policy Institute*, March 14, 2019. energyand-policy.org/green-new-deal-cost.

Angus Reid Institute. 2021. *Climate Change: O'Toole's Carbon Pricing Gamble Draws Mixed Political Reviews*. angusreid.org/wp-content/up-loads/2021/05/2021.05.03_Climate_Change.pdf.

Angus, Ian. 2016. *Facing the Anthropocene: Fossil Capitalism and the Crisis of the Earth System*. New York: Monthly Review Press.

AP-NORC. (The Associated Press-NORC Center for Public Affairs Research). 2022. *Immigration Attitudes and Conspiratorial Thinkers: A Study Issued on the 10th Anniversary of The Associated Press-NORC Center for Public Affairs Research*. apnorc.org/projects/immigration-attitudes-and-conspira-torial-thinkers.

Arabella Advisors. 2015. *Measuring the Growth of the Global Fossil Fuel Divestment and Clean Energy Investment Movement*. arabellaadvisors.com/wp-content/uploads/2016/10/Measuring-the-Growth-of-the-Divestment-Movement.pdf.

Aronoff, Kate. 2019. "The European Far Right's Environmental Turn." *Dissent*, May 31, 2019. dissentmagazine.org/online_articles/the-europe-an-far-rights-environmental-turn.

Aronoff, Kate, Alyssa Battistoni, Daniel Aldana Cohen, and Thea Riofrancos. 2019. *A Planet to Win: Why We Need a Green New Deal*. London: Verso. EPUB.

Atkin, Emily. 2019. "The Potency of Republicans' Hamburger Lie." *New Republic*, March 4, 2019. newrepublic.com/article/153187/potency-repub-licans-hamburger-lie.

____. 2020. "Facebook Creates Fact-Checking Exemption for Climate Deniers."

Heated, June 24, 2020. heated.world/p/facebook-creates-fact-checking-exemption.

___. 2021a. "The True Cost of Fossil Fuel Subsidies." *Heated*, April 6, 2021. heated.world/p/the-true-cost-of-fossil-fuel-subsidies.

___. 2021b. "Twitter's Big Oil Ad Loophole." *Heated*, February 28, 2021. heated.world/p/twitters-big-oil-ad-loophole.

Avaaz. 2020. *Why Is YouTube Broadcasting Climate Misinformation to Millions?* avaazimages.avaaz.org/youtube_climate_misinformation_report.pdf.

Bacon, Wendy, and Arunn Jegan. 2020. *Lies, Debates, and Silences: How News Corp Produces Climate Scepticism in Australia.* GetUp. d68ej2dhhub09.cloudfront.net/2790-Lies_Debates_and_Silences_FINAL.pdf.

Baer, Paul, Tom Athanasiou, Sivan Kartha, and Eric Kemp-Benedict. 2008. *The Greenhouse Development Rights Framework: The Right to Development in a Climate Constrained World.* Berlin: Heinrich Böll Foundation.

Bales, Kevin. 2016. *Blood and Earth: Modern Slavery, Ecocide, and the Secret to Saving the World.* New York: Spiegel & Grau.

Ball, Terrence. 2006. "Democracy." In *Political Theory and the Ecological Challenge*, edited by Andrew Dobson and Robyn Eckersley. New York: Cambridge University Press.

Bárány, Ambrus, and Dalia Grigonytė. 2015. *Measuring Fossil Fuel Subsidies.* ECFIN Economic Brief 40. ec.europa.eu/economy_finance/publications/economic_briefs/2015/pdf/eb40_en.pdf.

Barthold, Corbin K. 2020. "(Still) Against Degrowth." *Forbes*, April 29, 2020. forbes.com/sites/wlf/2020/04/29/still-against-degrowth.

Basden, Stuart. 2019. "Extinction Rebellion Isn't about the Climate." *Medium*, January 10, 2019. medium.com/extinction-rebellion/extinction-rebellion-isnt-about-the-climate-42a0a73d9d49.

Bastani, Aaron. 2019. *Fully Automated Luxury Communism: A Manifesto.* London: Verso.

Bayrak, Mucahid Mustafa, and Lawal Mohammed Marafa. 2016. "Ten Years of REDD+: A Critical Review of the Impact of REDD+ on Forest-Dependent Communities." *Sustainability* 8, no. 7. https://doi.org/10.3390/su8070620.

Beauchamp, Zack. 2019. " 'He's Not Hurting the People He Needs to Be': A Trump Voter Says the Quiet Part Out Loud." *Vox*, January 8, 2019. vox.com/policy-and-politics/2019/1/8/18173678/trump-shutdown-voter-florida.

Berardelli, Jeff. 2020. "How Joe Biden's Climate Plan Compares to the Green New Deal." CBS *News*, October 5, 2020. cbsnews.com/news/green-new-deal-joe-biden-climate-change-plan.

Berman, Sheri. 2020. "Can Democrats Save the World (Again)?" *Foreign Policy*, January 15, 2020. foreignpolicy.com/2020/01/15/social-democracy-save-world-again-socialism.

Black, Toban, Stephen D'Arcy, Tony Weis, and Joshua Kahn Russell, eds. 2014. *A Line in the Tar Sands: Struggles for Environmental Justice.* Oakland, CA: PM Press.

Blowe, Kevin. 2019. "It Is Not Just a Bunch of Flowers." *Medium*, October 16, 2019. copwatcher-uk.medium.com/it-is-not-just-a-bunch-of-flowers-bc5078b899e4.

Blum, William. 2004. *Killing Hope: U.S. Military and CIA Interventions Since World War II*. Monroe, ME: Common Courage Press.

Bookchin, Murray. 1989. *Remaking Society*. Montreal: Black Rose Books.

Bozuwa, Johanna, and Gar Alperovitz. 2019. "Electric Companies Won't Go Green Unless the Public Takes Control." *In These Times*, April 22, 2019. inthesetimes.com/article/electricity-renewable-energy-green-new-deal.

BP. 2021. *BP Statistical Review of World Energy 2021*. bp.com/content/dam/bp/business-sites/en/global/corporate/pdfs/energy-economics/statistical-review/bp-stats-review-2021-full-report.pdf.

Brand, Ulrich, and Markus Wissen. 2021. *The Imperial Mode of Living: Everyday Life and the Ecological Crisis of Capitalism*. London: Verso.

Braungart, Michael, and William McDonough. 2002. *Cradle to Cradle: Remaking the Way We Make Things*. New York: North Point Press.

Brecher, Jeremy. 2020a. "Black Lives and the Green New Deal." *CommonDreams*, June 11, 2020. commondreams.org/views/2020/06/11/black-lives-and-green-new-deal.

____. 2020b. "No Worker Left Behind: Protecting Workers and Communities in the Green New Deal." *New Labor Forum* 29, no. 2.

Bregman, Rutger. 2017. *Utopia for Realists: How We Can Build the Ideal World*. Translated by Elizabeth Manton. New York: Little, Brown and Company.

Brown, Tisha. 2015. "DeС02lonalism 101: We Need to Talk about Oppression." *New Internationalist*, December 2, 2015. newint.org/blog/guests/2015/12/02/we-need-to-talk-about-oppression.

Brown, Wendy. 2015. *Undoing the Demos: Neoliberalism's Stealth Revolution*. Brooklyn, NY: Zone Books.

Bruggeman, Lucien, Devin Dwyer, and Stephanie Ebbs. 2022. "Climate Activist's Fight against 'Terrorism' Sentence Could Impact the Future of Protests." *ABC News*, April 28, 2022. abcnews.go.com/US/climate-activists-fight-terrorism-sentence-impact-future-protests/story?id=84345514.

Brulle, Robert J. 2018. "The Climate Lobby: A Sectoral Analysis of Lobbying Spending on Climate Change in the USA, 2000 to 2016." *Climatic Change* 149. https://doi.org/10.1007/s10584-018-2241-z.

Buck, Holly Jean. 2019. *The World after Geoengineering: Climate Tragedy, Repair, and Restoration*. London: Verso.

____. 2020. "Should Carbon Removal Be Treated as Waste Management? Lessons from the Cultural History of Waste." *Interface Focus* 10. https://doi.org/10.1098/rsfs.2020.0010.

Bump, Philip. 2019. "It's Possible that Trump Doesn't Actually Know What Climate Change Is." *Washington Post*, December 3, 2019. washingtonpost.com/politics/2019/12/03/its-possible-that-trump-doesnt-actually-know-what-climate-change-is.

Cames, Martin, Ralph O. Harthan, Jürg Füssler, et al. 2016. *How Additional Is the Clean Development Mechanism? Analysis of the Application of Current Tools and Proposed Alternatives.* European Commission. ec.europa.eu/clima/sites/clima/files/ets/docs/clean_dev_mechanism_en.pdf.

Carapella, Aaron. 2016. *Proposed Pipelines in Tribal Homelands.* tribalnations-maps.com/pipeline-map.html.

Carlson, Tucker. 2021. "Tucker Carlson Goes One-on-One with Outkick." Interview by Bobby Burack. *Outkick*, April 23, 2021. outkick.com/tucker-carlson-fox-news.

Carter, Lawrence. 2021. "Inside Exxon's Playbook: How America's Biggest Oil Company Continues to Oppose Action on Climate Change." *Unearthed*, June 30, 2021. unearthed.greenpeace.org/2021/06/30/exxon-climate-change-undercover.

Cave, Damien. 2020. "How Rupert Murdoch Is Influencing Australia's Bushfire Debate." *New York Times*, January 13, 2020. nytimes.com/2020/01/08/world/australia/fires-murdoch-disinformation.html.

CBC News. 2014a. "Clean Energy Provides More Jobs than Oilsands, Report Says." CBC, December 2, 2014. cbc.ca/news/business/clean-energy-provides-more-jobs-than-oilsands-report-says-1.2857520.

____. 2014b. "Tony Abbott, Stephen Harper Take Hard Line against Carbon Tax." June 9, 2014. Video, 2:50. cbc.ca/player/play/2463534279.

Chait, Jonathan. 2015. "Is Naomi Klein Right that We Must Choose between Capitalism and the Climate?" *Intelligencer*, October 23, 2015. nymag.com/intelligencer/2015/10/must-we-choose-between-capitalism-and-climate.html.

____. 2019. "The Green New Deal Is a Bad Idea, Not Just a Botched Rollout." *Intelligencer*, February 12, 2019. nymag.com/intelligencer/2019/02/green-new-deal-aoc-bad-idea.html.

Chancel, Lucas, and Thomas Piketty. 2015. *Carbon and Inequality: From Kyoto to Paris.* Paris: Paris School of Economics.

Chancel, Lucas, Thomas Piketty, Emmanuel Saez, et al. 2021. *World Inequality Report 2022.* World Inequality Lab. wir2022.wid.world.

Chomsky, Noam. 2003. *Hegemony or Survival: America's Quest for Global Dominance.* New York: Holt Paperbacks.

____. 2011. "Noam Chomsky: Anarchism, Council Communism, and Life After Capitalism." Interview by Sasha Lilley. In *Capital and Its Discontents: Conversations with Radical Thinkers in a Time of Tumult*, edited by Sasha Lilley. Oakland, CA: PM Press.

____. 2013. "The Kind of Anarchism I Believe In, and What's Wrong with Libertarians." Interviewed by Michael S. Wilson. *chomsky.info*, May 28, 2013. chomsky.info/20130528.

____. 2020. "Noam Chomsky's Green New Deal." Interview by David Roberts. *Vox*, September 21, 2020. vox.com/energy-and-environment/21446383/noam-chomsky-robert-pollin-climate-change-book-green-new-deal.

Chomsky, Noam, Peter Hutchison, Kelly Nyks, and Jared P. Scott, eds. 2017.

Requiem for the American Dream: The 10 Principles of Concentration of Wealth & Power. New York: Seven Stories Press.

Churchill, Ward. 2007. *Pacifism as Pathology: Reflections on the Role of Armed Struggle in North America.* Oakland, CA: AK Press.

Ciplet, David. 2017. "Subverting the Status Quo? Climate Debt, Vulnerability and Counter-Hegemonic Frame Integration in United Nations Climate Politics – A Framework for Analysis." *Review of International Political Economy* 24, no. 6. https://doi.org/10.1080/09692290.2017.1392336.

Clausing, Kimberly, Emmanuel Saez, and Gabriel Zucman. 2021. *Ending Corporate Tax Avoidance and Tax Competition: A Plan to Collect the Tax Deficit of Multinationals.* UCLA School of Law, Law-Econ Research Paper No. 20-12. http://doi.org/10.2139/ssrn.3655850.

Climate Action Tracker. 2021. *Glasgow's One Degree 2030 Credibility Gap: Net Zero's Lip Service to Climate Action.* New Climate Institute and Climate Analytics. climateactiontracker.org/press/Glasgows-one-degree-2030-credibility-gap-net-zeros-lip-service-to-climate-action.

Climate Disobedience Center. n.d. "The Climate Necessity Defense: A Legal Tool for Climate Activists." climatedisobedience.org/necessitydefense.

Climate Leadership Council. 2019. *Economists' Statement on Carbon Dividends.* clcouncil.org/economists-statement.

Coady, David, Ian Parry, Louis Sears, and Baoding Shang. (2015). *How Large Are Global Energy Subsidies?* IMF Working Paper. imf.org/external/pubs/ft/wp/2015/wp15105.pdf.

Coan, Travis G., Constantine Boussalis, John Cook, and Mirjam O. Nanko. 2021. "Computer-Assisted Classification of Contrarian Claims about Climate Change." *Nature* 11. https://doi.org/10.1038/s41598-021-01714-4.

Coleman, Aaron Ross. 2019. "How Black Lives Matter to the Green New Deal." *Nation,* March 14, 2019. thenation.com/article/archive/reparations-green-new-deal-aoc.

Colman, Zack. 2019. "The Bogus Number at the Center of the GOP's Green New Deal Attacks." *Politico,* March 10, 2019. politico.com/story/2019/03/10/republican-green-new-deal-attack-1250859.

Confessore, Nicholas. 2022. "How Tucker Carlson Stoked White Fear to Conquer Cable." *New York Times,* April 30, 2022. nytimes.com/2022/04/30/us/tucker-carlson-gop-republican-party.html.

Cook, John. 2017. "The Quantum Theory of Climate Denial." *HuffPost,* December 6, 2017. huffpost.com/entry/the-quantum-theory-of-climate-de-nial_b_5229539.

____. 2020. "A History of FLICC: The 5 Techniques of Science Denial." *Cranky Uncle,* March 24, 2020. crankyuncle.com/a-history-of-flicc-the-5-tech-niques-of-science-denial.

Cook, John, Geoffrey Supran, Stephan Lewandowsky, et al. 2019. *America Misled: How the Fossil Fuel Industry Deliberately Misled Americans about Climate Change.* Fairfax, VA: George Mason University Center for Climate Change

Communication. climatechangecommunication.org/america-misled.

Cosme, Inês, Rui Santos, and Daniel W. O'Neill. 2017. "Assessing the Degrowth Discourse: A Review and Analysis of Academic Degrowth Policy Proposals." *Journal of Cleaner Production* 149. http://doi.org/10.1016/j.jclepro.2017.02.016.

Cox, Stan. 2020. *The Green New Deal and Beyond: Ending the Climate Emergency While We Still Can.* San Francisco: City Lights Books.

Coyne, Andrew. 2019. "Andrew Coyne: Parties' Climate Change Policies Range from the Inadequate to the Insane." *National Post*, January 31, 2019. nationalpost.com/opinion/andrew-coyne-parties-climate-change-policies-range-from-the-inadequate-to-the-insane.

Cranley, Ellen. 2020. "Rupert Murdoch's News Corp Criticized for Misleading Bushfire Coverage Amplified by Trolls and Bots." *Insider*, January 11, 2020. insider.com/rupert-murdochs-news-corp-australia-bushfires-2020-1.

Credit Suisse. 2020. *Global Wealth Report 2020.* credit-suisse.com/about-us/en/reports-research/global-wealth-report.html.

Crenshaw, Dan. 2020. "It's Time for Conservatives to Own the Climate-Change Issue." *National Review*, March 3, 2020. nationalreview.com/2020/03/its-time-for-conservatives-to-own-the-climate-change-issue.

Dale, Gareth. 2019. "Degrowth and the Green New Deal." *Ecologist*, October 28, 2019. theecologist.org/2019/oct/28/degrowth-and-green-new-deal.

Daly, Herman E. 2005. "Economics in a Full World." *Scientific American*, September 1, 2005. scientificamerican.com/article/economics-in-a-full-world.

____. 2014. *From Uneconomic Growth to a Steady-State Economy.* Cheltenham, UK: Edward Elgar. https://doi.org/10.4337/9781783479979.

____. 2015. "Economics for a Full World." *Great Transition Initiative*, June 2015. greattransition.org/publication/economics-for-a-full-world.

Dauvergne, Peter. 2008. *The Shadows of Consumption: Consequences for the Global Environment.* Cambridge, MA: MIT Press.

____. 2016. *Environmentalism of the Rich.* Cambridge, MA: MIT Press.

Delina, Laurence L. 2016. *Strategies for Rapid Climate Mitigation.* New York: Routledge.

Delina, Laurence L., and Mark Diesendorf. 2013. "Is Wartime Mobilisation a Suitable Policy Model for Rapid National Climate Mitigation?" *Energy Policy* 58. https://doi.org/10.1016/j.enpol.2013.03.036.

de Ferrer, Marthe. 2021. "GB News Criticised for Platforming 'Dangerous Climate Change Deniers.' " *EuroNews*, June 29 2021. euronews.com/green/2021/06/26/gb-news-criticised-for-platforming-dangerous-climate-change-deniers.

della Porta, Donatella, and Louisa Parks. 2014. "Framing Processes in the Climate Movement: From Climate Change to Climate Justice." In *Routledge Handbook of the Climate Movement,* edited by M. Dietz and H. Garrelts. New York: Routledge.

de Moor, Joost, Katrin Uba, Mattias Wahlström, et al. 2020. "Introduction: Fridays for Future – An Expanding Climate Movement." In *Protest for a Future II: Composition, Mobilization and Motives of the Participants in Fridays for Future Climate Protests on 20-27 September, 2019, in 19 Cities around the World*, edited by Joost de Moor, Katrin Uba, Mattias Wahlström, et al. osf.io/yg9k2.

Dhillon, Jaskiran, and Will Parrish. 2019. "Exclusive: Canada Police Prepared to Shoot Indigenous Activists, Documents Show." *The Guardian*, December 20, 2019. theguardian.com/world/2019/dec/20/canada-indigenous-land-defenders-police-documents.

Diffenbaugh, Noah S., and Marshall Burke. 2019. "Global Warming Has Increased Global Economic Inequality." *Proceedings of the National Academy of Sciences of the United States of America* 116, no. 20. https://doi.org/10.1073/pnas.1816020116.

Dobson, Andrew. 2013. "Political Theory in a Closed World: Reflections on William Ophuls, Liberalism and Abundance." *Environmental Values* 22, no. 2. https://doi.org/10.3197/096327113X13581561725275.

Dodd, Vikram, and Jamie Grierson. 2020. "Terrorism Police List Extinction Rebellion as Extremist Ideology." *The Guardian*, January 10, 2020. theguardian.com/uk-news/2020/jan/10/xr-extinction-rebellion-listed-extremist-ideology-police-prevent-scheme-guidance.

Donohue, Thomas J. 2019. "The Green New Deal Is a Trojan Horse for Socialism." *U.S. Chamber of Commerce*, February 18, 2019. uschamber.com/series/above-the-fold/the-green-new-deal-trojan-horse-socialism.

Dorninger, Christian, Alf Hornborg, David J. Abson, et al. 2021. "Global Patterns of Ecologically Unequal Exchange: Implications for Sustainability in the 21st Century." *Ecological Economics* 179. https://doi.org/10.1016/j.ecolecon.2020.106824.

Draper, Hal. 1966. *The Two Souls of Socialism*. Berkeley, CA: Independent Socialist Committee.

Drennen, Ari, and Sally Hardin. 2021. "Climate Deniers in the 117th Congress." *Center for American Progress*, March 30, 2021. americanprogress.org/issues/green/news/2021/03/30/497685/climate-deniers-117th-congress.

Druckman, Angela, Ian Buck, Bronwyn Hayward, and Tim Jackson. 2012. "Time, Gender and Carbon: A Study of the Carbon Implications of British Adults' Use of Time." *Ecological Economics* 84. https://doi.org/10.1016/j.ecolecon.2012.09.008.

DSA Ecosocialist Working Group. 2019. "DSA's Green New Deal Principles." February 28, 2019. ecosocialists.dsausa.org/2019/02/28/gnd-principles.

Dunlap, Riley E. 2008. "Partisan Gap on Global Warming Grows." *Gallup*, May 29, 2008. news.gallup.com/poll/107593/partisan-gap-global-warming-grows.aspx.

Dunlap, Riley E., and Aaron M. McCright. 2011. "Organized Climate Change Denial." In *The Oxford Handbook of Climate Change and Society*, edited by

John S. Dryzek, Richard B. Norgaard, and David Schlosberg. Oxford, UK: Oxford University Press.

Earth Overshoot Day. 2022. "This Year's Earth Overshoot Day Falls on July 28." News release, June 5, 2022. overshootday.org/newsroom/press-release-june-2022-english.

Eckersley, Robyn. 2004. *The Green State: Rethinking Democracy and Sovereignty*. Cambridge, MA: MIT Press.

Ellingboe, Kristen, Tiffany Germain, and Kiley Kroh. 2015. "The Anti-Science Climate Denier Caucus: 114th Congress Edition." *ThinkProgress*, January 8, 2015. archive.thinkprogress.org/the-anti-science-climate-denier-caucus-114th-congress-edition-c76c3f8bfedd/.

Evershed, Nick. 2020. "Australia's Newspaper Ownership Is among the Most Concentrated in The World." *The Guardian*, November 13, 2020. theguardian.com/news/datablog/2020/nov/13/australia-newspaper-ownership-is-among-the-most-concentrated-in-the-world.

Extinction Rebellion. 2020a. "BREAKING: Extinction Rebellion Blocks News Corps Printworks and Demands They 'Free the Truth.' " September 4, 2020. extinctionrebellion.uk/2020/09/04/breaking-extinction-rebellion-blocks-news-corps-printworks-and-demands-they-free-the-truth.

____. 2020b. "Statement on Extinction Rebellion's Relationship with the Police." July 1, 2020. extinctionrebellion.uk/2020/07/01/statement-on-extinction-rebellions-relationship-with-the-police.

Farand, Chloé. 2020. "US Climate Activists Confront the Movement's Whiteness Problem." *Climate Home News*, June 23, 2020. climatechangenews.com/2020/06/23/us-climate-activists-confront-movements-whiteness-problem.

Farand, Chloé, Maribel Ángel-Moreno, Léopold Salzenstein, and Jelena Malkowski. 2022. "Data Exclusive: The 'Junk' Carbon Offsets Revived by the Glasgow Pact." *Climate Home News*, June 17, 2022. climatechangenews.com/2022/06/17/data-exclusive-the-junk-carbon-offsets-revived-by-the-glasgow-pact.

Fleming, Amy. 2021. "Cloud Spraying and Hurricane Slaying: How Ocean Geoengineering Became the Frontier of the Climate Crisis." *The Guardian*, June 23, 2021. theguardian.com/environment/2021/jun/23/cloud-spraying-and-hurricane-slaying-could-geoengineering-fix-the-climate-crisis.

Folke, Carl, Stephen Polasky, Johan Rockström, et al. 2021. "Our Future in the Anthropocene Biosphere." *Ambio* 50. https://doi.org/10.1007/s13280-021-01544-8.

Freeden, Michael. 2003. *Ideology: A Very Short Introduction*. Oxford, UK: Oxford University Press.

Fremstad, Anders, and Mark Paul. 2022. "Neoliberalism and Climate Change: How the Free-Market Myth Has Prevented Climate Action." *Ecological Economics* 197. https://doi.org/10.1016/j.ecolecon.2022.107353.

Friedman, Thomas L. 2009. *Hot, Flat, and Crowded: Why We Need a Green*

Revolution and How It Can Renew America (Release 2.0: Updated and Expanded). Toronto: Douglas & McIntyre.

Funk, Cary, Alec Tyson, Brian Kenney, and Courtney Johnson. 2020. "Science and Scientists Held in High Esteem Across Global Publics." *Pew Research Center*, September 29, 2020. pewresearch.org/science/2020/09/29/concern-over-climate-and-the-environment-predominates-among-these-publics.

Gallup. 2022. "Gallup Poll Social Series: Environment." *Gallup*. news.gallup.com/file/poll/391520/220406ExtremeWeather.pdf.

Gardiner, Stephen M. 2011. *A Perfect Moral Storm: The Ethical Tragedy of Climate Change*. Oxford, UK: Oxford University Press.

Gayle, Damien. 2019. "Does Extinction Rebellion Have a Race Problem?" *The Guardian*, October 4, 2019. theguardian.com/environment/2019/oct/04/extinction-rebellion-race-climate-crisis-inequality.

Gebriel, Dalia. 2019. "As the Left Wakes Up to Climate Injustice, We Must Not Fall into 'Green Colonialism.' " *The Guardian*, May 8, 2019. theguardian.com/commentisfree/2019/may/08/left-climate-injustice-green-new-deal.

Giannetti, B.F., F. Agostinho, C.M.V.B. Almeida, and D. Huisingh. 2015. "A Review of Limitations of GDP and Alternative Indices to Monitor Human Wellbeing and to Manage Eco-System Functionality." *Journal of Cleaner Production* 87. http://doi.org/10.1016/j.jclepro.2014.10.051.

Gilbertson, Tamra. 2017. *Carbon Pricing: A Critical Perspective for Community Resistance*. Bemidji, MN: Indigenous Environmental Network and Climate Justice Alliance. co2colonialism.org/wp-content/uploads/2019/11/Carbon-Pricing-A-Critical-Perspective-for-Community-Resistance-Online-Version.pdf.

Gilbertson, Tamra, and Oscar Reyes. 2009. *Carbon Trading: How It Works and Why It Fails*. Uppsala, Sweden: Dag Hammarskjöld Foundation.

Givens, Jennifer E., Xiaorui Huang, and Andrew K. Jorgenson. 2019. "Ecologically Unequal Exchange: A Theory of Global Environmental Injustice." *Sociology Compass* 13. https://doi.org/10.1111/soc4.12693.

Global Footprint Network. 2022. "Country Trends." data.footprintnetwork.org/#/countryTrends.

Goldtooth, Tom B.K. 2011a. "What Does Climate Change Mean for Indigenous Communities?" YouTube video, 7:03, uploaded December 27, 2011 by "One World TV." youtube.com/watch?v=PFRxJFUefw8.

___. 2011b. "Why REDD/REDD+ is NOT a Solution." In *No REDD Papers Volume One*, edited by Hallie Boas. Portland, OR: Charles Overbeck/Eberhardt Press. ienearth.org/docs/No-Redd-Papers.pdf.

Gore, Al. 2009. *Our Choice: A Plan to Solve the Climate Crisis*. Emmaus, PA: Rodale Books.

Gorz, André. 2012. *Capitalism, Socialism, Ecology*. Translated by Martin Chalmers. London: Verso. First published 1994.

Graziosi, Graig. 2020. "Tucker Carlson Says Climate Change Is a Liberal Invention 'Like Racism' in Shocking On-Air Rant." *Independent*,

September 13, 2020. independent.co.uk/news/world/americas/tucker-carlson-climate-change-fox-news-california-wildfires-racism-liberal-b434261.html.

Greenpeace. n.d. "Koch Industries: Secretly Funding the Climate Denial Machine." https://www.greenpeace.org/usa/fighting-climate-chaos/climate-deniers/koch-industries.

Guardian. 2020. "Five Ways to Make the Climate Movement Less White." *The Guardian*, September 21, 2020. theguardian.com/us-news/2020/sep/21/five-ways-to-make-the-climate-movement-less-white.

Gunn-Wright, Rhiana. 2020. "Policies and Principles of a Green New Deal." In *Winning the Green New Deal: Why We Must, How We Can*, edited by Varshini Prakash and Guido Girgenti. New York: Simon and Schuster.

Gunn-Wright, Rhiana, and Robert Hockett. 2019. *The Green New Deal: Mobilizing for a Just, Prosperous, and Sustainable Economy*. New Consensus. https://newconsensus.com/files/gnd-overview.pdf.

Gustafson, A., Leiserowitz, A., Maibach, E. W., Rosenthal, S. A., Kotcher, J. K., and Goldberg, M. H. 2020. *Climate Change in the Minds of U.S. Media Audiences*. New Haven, CT: Yale Program on Climate Change Communication.

Hadden, Jennifer. 2014. "Explaining Variations in Transnational Climate Change Activism: The Role of Inter-Movement Spillover." *Global Environmental Politics* 14. https://doi.org/10.1162/GLEP_a_00225.

Hagens, N.J. 2020. "Economics for the Future – Beyond the Superorganism." *Ecological Economics* 169. https://doi.org/10.1016/j.ecolecon.2019.106520.

Hahnel, Robin, and Erik Olin Wright. 2016. *Alternatives to Capitalism: Proposals for a Democratic Economy*. Brooklyn, NY: Verso.

Halon, Yael. 2021. "Bill Gates Support for 'Bonkers' Study of Dimming the Sun Is 'Grossly Irresponsible:' Author." *Fox News*, April 8, 2021. foxnews.com/media/bill-gates-backs-project-to-dim-the-sun-michael-shellenberger.

Hamilton, Clive. 2004. *Growth Fetish*. London: Pluto Press.

____. 2010. *Requiem for a Species: Why We Resist the Truth about Climate Change* Washington, DC: Earthscan.

____. 2013. *Earthmasters: The Dawn of the Age of Climate Engineering*. New Haven, CT: Yale University Press.

____. 2017. *Defiant Earth: The Fate of Humans in the Anthropocene*. Malden, MA: Polity.

Hansen, James and Makiko Sato. 2021. "Fossil Fuel CO2 Emissions." columbia.edu/~mhs119/CO2Emissions/Emis_moreFigs/.

Hardin, Sally, and Claire Moser. 2019. "Climate Deniers in the 116th Congress." *Center for American Progress Action Fund*, January 28, 2019. americanprogressaction.org/issues/green/news/2019/01/28/172944/climate-deniers-116th-congress.

Harvey, David. 2003. *The New Imperialism*. Oxford, UK: Oxford University Press.

___. 2005. *A Brief History of Neoliberalism*. Oxford, UK: Oxford University Press.

___. 2009. *Cosmopolitanism and the Geographies of Freedom*. New York: Columbia University Press.

___. 2014. *Seventeen Contradictions and the End of Capitalism*. New York: Oxford University Press.

Hawkin, Paul, Amory Lovins, and L. Hunter Lovins. 2000. *Natural Capitalism: Creating the Next Industrial Revolution*. Washington, DC: US Green Building Council.

Heglar, Mary Annaïse, and Amy Westervelt. 2020. "Yes, It's Still Time to Talk About Climate." *Hot Take*, June 5, 2020. Audio, 26:11. podbay.fm/p/1488414960/e/1591356600.

Helliwell, John F., Richard Layard, Jeffrey Sachs, and Jan-Emmanuel De Neve, eds. 2020. *World Happiness Report 2020*. New York: Sustainable Development Solutions Network. worldhappiness.report/ed/2020.

Hermes, Karin Louise. 2020. "Why I Quit Being a Climate Activist." *Vice*, February 6, 2020. vice.com/en_us/article/g5x5ny/why-i-quit-being-a-climate-activist.

Hess, Abigail Johnson. 2020. "As People Protest across the U.S., Some Wonder: Could You Be Fired for Protesting?" *CNBC Make It*, June 5, 2020. cnbc.com/2020/06/05/can-you-get-fired-for-attending-a-protest.html.

Hickel, Jason. 2020a. "Just to be clear: the economic contraction that's happening right now is *not* degrowth." *Twitter*, April 29, 2020, 4:08 p.m. twitter.com/jasonhickel/status/1255589713915908096.

___. 2020b. *Less Is More: How Degrowth Will Save the World*. London: William Heineman.

___. 2020c. "Quantifying National Responsibility for Climate Breakdown: An Equality-Based Attribution Approach for Carbon Dioxide Emissions in Excess of the Planetary Boundary." *Lancet Planet Health* 4. https://doi.org/10.1016/S2542-5196(20)30196-0.

___. 2020d. "The Sustainable Development Index: Measuring the Ecological Efficiency of Human Development in the Anthropocene." *Ecological Economics* 167. https://doi.org/10.1016/j.ecolecon.2019.05.011.

Hickel, Jason, Christian Dorninger, Hanspeter Wieland, and Intan Suwandi. 2022. "Imperialist Appropriation in the World Economy: Drain from the Global South through Unequal Exchange, 1990–2015." *Global Environmental Change* 73. https://doi.org/10.1016/j.gloenvcha.2022.102467.

High-Level Commission on Carbon Prices. 2017. *Report of the High-Level Commission on Carbon Prices*. Washington, DC: World Bank.

Hochschild, Arlie Russell. 2016. *Strangers in Their Own Land: Anger and Mourning on the American Right*. New York: The New Press. EPUB.

Homer-Dixon, Thomas. 2016. "Start the Leap Revolution without Me." *Globe and Mail*, April 22, 2016. theglobeandmail.com/opinion/start-the-leap-revolution-without-me/article29711945.

Hornsey, Matthew J., Emily A. Harris, Paul G. Bain, and Kelly S. Fielding. 2016.

"Meta-Analyses of the Determinants and Outcomes of Belief in Climate Change." *Nature Climate Change* 6. https://doi.org/10.1038/nclimate2943.

Hourdequin, Marion. 2018. "Climate Change, Climate Engineering, and the 'Global Poor': What Does Justice Require?" *Ethics, Policy & Environment* 21, no. 3. https://doi.org/10.1080/21550085.2018.1562525.

Huertas, Aaron, and Rachel Kriegsman. 2014. *Science or Spin? Assessing the Accuracy of Cable News Coverage of Climate Science.* Union of Concerned Scientists. ucsusa.org/sites/default/files/2019-09/Science-or-Spin-report.pdf.

Independent Expert Group on Climate Finance. 2020. *Delivering on the $100 Billion Climate Finance Commitment and Transforming Climate Finance.* https://www.un.org/sites/un2.un.org/files/2020/12/100_billion_climate_finance_report.pdf.

Indigenous Environmental Network and Oil Change International. 2021. *Indigenous Resistance Against Carbon.* Washington, DC: Oil Change International. ienearth.org/indigenous-resistance-against-carbon/.

InfluenceMap. 2020. *Climate Change and Digital Advertising: Climate Science Disinformation in Facebook Advertising.* influencemap.org/report/Climate-Change-and-Digital-Advertising-86222daed29c6f49ab2da76b0df15f76.

Intergovernmental Panel on Climate Change. 2021. "Summary for Policymakers." In *Climate Change 2021: The Physical Science Basis. Contribution of Working Group I to the Sixth Assessment Report of the Intergovernmental Panel on Climate Change,* edited by V. Masson-Delmotte, P. Zhai, A. Pirani, et al. Cambridge, UK: Cambridge University Press. https://www.doi.org/10.1017/9781009157896.001.

Iqbal, Nosheen. 2020. "Climate Activists Accused of 'Attacking Free Press' by Blockading Print Works." *The Guardian*, September 5, 2020. theguardian.com/environment/2020/sep/05/climate-activists-accused-of-attacking-free-press-by-blockading-print-works.

Jaccard, Mark. 2019. "If Canadians Elect a Climate-Insincere Government in 2019, Climate-Concerned Voters May Need to Look in the Mirror when Allocating Blame." *Sustainability Suspicions* (blog), August 1, 2019. markjaccard.blogspot.com/2019/08/if-canadians-elect-climate-insincere.html.

___. 2020. *The Citizen's Guide to Climate Success: Overcoming Myths That Hinder Progress.* Cambridge, UK: Cambridge University Press. PDF Book.

Jackson, Tim. 2009. *Prosperity Without Growth: Economics for a Finite Planet.* London: Earthscan.

Jafino, Bramka Arga, Brian Walsh, Julie Rozenberg, and Stephane Hallegatte. 2020. *Revised Estimates of the Impact of Climate Change on Extreme Poverty by 2030.* Policy Research Working Paper No. 9417. Washington, DC: World Bank. hdl.handle.net/10986/34555.

Jensen, Derrick. 2006a. *Endgame: Volume 1: The Problem of Civilization.* New York: Seven Stories Press.

___. 2006b. *Endgame: Volume 2: Resistance.* New York: Seven Stories Press.

Johnson, Brad. 2010. "The Climate Zombie Caucus of the 112th Congress."

ThinkProgress, November 19, 2010. archive.thinkprogress.org/the-climate-zombie-caucus-of-the-112th-congress-2ee9c4f9e46/.

Johnson, Jake. 2021. "'He Is Lying. People Are Dying': Calls for Texas Governor to Resign as He Blames Power Outages on Wind and Solar." *Common Dreams*, February 17, 2021. commondreams.org/news/2021/02/17/he-lying-people-are-dying-calls-texas-governor-resign-he-blames-power-outages-wind.

Kahan, Dan M., Hank Jenkins-Smith, and Donald Braman. 2011. "Cultural Cognition of Scientific Consensus." *Journal of Risk Research* 14, no. 2. https://doi.org/10.1080/13669877.2010.511246.

Kallis, Giorgos. 2019. "Capitalism, Socialism, Degrowth: A Rejoinder." *Capitalism Nature Socialism* 30, no. 2. https://doi.org/10.1080/10455752.2018.1563624.

Kallis, Giorgos, Susan Paulson, Giacomo D'Alisa, and Federico Demaria. 2020. *The Case for Degrowth*. Medford, MA: Polity.

Kartha, Sivan, Eric Kemp-Benedict, Emily Ghosh, and Anisha Nazareth. 2020. *The Carbon Inequality Era: An Assessment of the Global Distribution of Consumption Emissions among Individuals from 1990 to 2015 and Beyond*. Oxford, UK: Oxfam and Stockholm Environment Institute.

Kaufman, Alexander C. 2021. "How Arizona's Attorney General Is Weaponizing Climate Fears to Keep Out Immigrants." *HuffPost*, May 1, 2021. huffpost.com/entry/arizona-climate-lawsuit_n_60897a42e4b0b9042d8d6ae5.

Keith, David W. 2013. *A Case for Climate Engineering*. Cambridge, MA: MIT Press.

Kelbert, Alexandra Wanjiku, and Joshua Virasami. 2015. "Darkening the White Heart of the Climate Movement." *New Internationalist*, 3 December, 2015. newint.org/blog/guests/2015/12/01/darkening-the-white-heart-of-the-climate-movement.

Kelton, Stephanie. 2020. *The Deficit Myth: Modern Monetary Theory and the Birth of the People's Economy*. New York: Public Affairs.

Kelton, Stephanie, Andres Bernal, and Greg Carlock. 2018. "We Can Pay for A Green New Deal." *HuffPost*, November 30, 2018. huffpost.com/entry/opinion-green-new-deal-cost_n_5c0042b2e4b027f1097bda5b.

Keyßer, Lorenz T., and Manfred Lenzen. 2021. "1.5 °C degrowth scenarios suggest the need for new mitigation pathways." *Nature Communications* 12. https://doi.org/10.1038/s41467-021-22884-9.

Keynes, John Maynard. [1931] 2010. *Essays in Persuasion*. Basingstoke, UK: Palgrave Macmillan.

Klein, Naomi. 2007. *The Shock Doctrine: The Rise of Disaster Capitalism*. New York: Metropolitan Books.

____. 2009. "Climate Rage." *Rolling Stone*, November 12, 2009. rollingstone.com/politics/politics-news/climate-rage-193377.

____. 2010. "Addicted to Risk." TED talk, December 8, 2010. naomiklein.org/on-precaution.

____. 2014a. *This Changes Everything: Capitalism vs. the Climate*. Toronto:

Alfred A. Knopf.

___. 2014b. "No, We Don't Need to Ditch/Slay/Kill Capitalism Before We Can Fight Climate Change. But We Sure as Hell Need to Challenge It." *This Changes Everything*, September 27, 2014. thischangeseverything.org/no-we-dont-need-to-ditchslaykill-capitalism-before-we-can-fight-climate-change-but-we-sure-as-hell-need-to-challenge-it.

___. 2014c. "The People's Climate March: Meet the Next Movement of Movements." *This Changes Everything*, September 14, 2014. thischangeseverything.org/the-peoples-climate-march-meet-the-next-movement-of-movements.

___. 2014d. "Why #BlackLivesMatter Should Transform the Climate Debate." *The Nation*, December 12, 2014. thenation.com/article/archive/what-does-blacklivesmatter-have-do-climate-change.

___. 2019. *On Fire: The Burning Case for a Green New Deal*. Toronto: Alfred A. Knopf.

___. 2020. "6. The New Shock Doctrine: A Conversation with Naomi Klein." Interview by Grace Blakeley. *A World to Win*, September 23, 2020. Audio, 47:42. tribunemag.co.uk/2020/09/6-the-new-shock-doctrine-a-conversation-with-naomi-klein.

Klein, Seth. 2020. *A Good War: Mobilizing Canada for the Climate Emergency*. Toronto: ECW Press.

Kostigen, Thomas M. 2020. *Hacking Planet Earth: How Geoengineering Can Help Us Reimagine the Future*. New York: TarcherPerigee.

Kovel, Joel. 2002. *The Enemy of Nature: The End of Capitalism or the End of the World?* London: Zed Books.

Krieg, Gregory, and Ella Nilsen. 2022. "An 'Excruciating Year': Climate Activists Reset with Biden's Agenda on Life Support." *CNN*, March 13, 2022. cnn.com/2022/03/13/politics/biden-climate-agenda-activists/index.html.

Lamb, William F., Giulio Mattioli, Sebastian Levi, et al. 2020. "Discourses of Climate Delay." *Global Sustainability* 3. https://doi.org/10.1017/sus.2020.13.

Latouche, Serge. 2009. *Farewell to Growth*. Malden, MA: Polity.

___. 2012. "Can the Left Escape Economism?" *Capital Nature Socialism* 23. https://doi.org/10.1080/10455752.2011.648841.

Lawrence, Mark G., Stefan Schäfer, Helene Muri, et al. 2018. "Evaluating Climate Geoengineering Proposals in the Context of the Paris Agreement Temperature Goals." *Nature Communications* 9, no. 3734. https://doi.org/10.1038/s41467-018-05938-3.

Lawrence, Mathew, and Laurie Laybourn-Langton. 2021. *Planet on Fire: A Manifesto for the Age of Environmental Breakdown*. Brooklyn, NY: Verso.

Lemire, Jonathan, Aamer Madhani, Will Weissert, and Ellen Knickmeyer. 2020. "Trump Spurns Science on Climate: 'Don't Think Science Knows.'" *ap News*, September 14, 2020. apnews.com/article/bd152cd786b58e45c61beb-f2457f9930.

Lewandowsky, Stephan, John Cook, and Elisabeth Lloyd. 2016. "The 'Alice in

Wonderland' Mechanics of the Rejection of (Climate) Science: Simulating Coherence by Conspiracism." *Synthese* 195. https://doi.org/10.1007/s11229-016-1198-6.

Lilac. 2013. "Deep Green Transphobia III: Derrick Jensen's Hateful Tirade." *EarthFirst! NewsWire*, May 17, 2013. earthfirstnews.wordpress.com/2013/05/17/deep-green-transphobia-iii-derrick-jensens-hateful-tirade.

Linnitt, Carol. 2015. "LEAKED: Internal RCMP Document Names 'Violent Anti-Petroleum Extremists' Threat to Government and Industry." *The Narwhal*, February 17, 2015. thenarwhal.ca/leaked-internal-rcmp-document-names-anti-petroleum-extremists-threat-government-industry.

Loewenstein, Antony. 2015. *Disaster Capitalism: Making a Killing Out of Catastrophe*. London: Verso.

Lynas, Mark, Benjamin Z. Houlton, and Simon Perry. 2021. "Greater than 99% Consensus on Human Caused Climate Change in the Peer-Reviewed Scientific Literature." *Environmental Research Letters* 16, no. 11. https://doi.org/10.1088/1748-9326/ac2966.

MacDonald, Ted. 2020. "Fox Breaks Out All the Tools in the Denier Playbook to Downplay Climate Change's Role in Western Wildfires." *Media Matters*, September 17, 2020. mediamatters.org/fox-news/fox-breaks-out-all-tools-denier-playbook-downplay-climate-changes-role-western-wildfires.

____. 2021. "How Broadcast TV Networks Covered Climate Change in 2020." *Media Matters*, March 10, 2021. mediamatters.org/broadcast-networks/how-broadcast-tv-networks-covered-climate-change-2020.

MacWilliams, Matthew C. 2020. "Trump Is an Authoritarian. So Are Millions of Americans." *Politico*, September 23, 2020. politico.com/news/magazine/2020/09/23/trump-america-authoritarianism-420681.

Magdoff, Fred. 2014. "An Ecologically Sound and Socially Just Economy." *Monthly Review* 66, no. 4. https://doi.org/10.14452/MR-066-04-2014-08_3.

Magdoff, Fred, and John Bellamy Foster. 2011. *What Every Environmentalist Needs to Know About Capitalism*. New York: Monthly Review Press.

Magdoff, Fred, and Chris Williams. 2017. *Creating an Ecological Society: Toward a Revolutionary Transformation*. New York: Monthly Review Press. EPUB.

Malm, Andreas. 2016a. *Fossil Capital: The Rise of Steam Power and the Roots of Global Warming*. London: Verso.

____. 2016b. "Who Lit This Fire? Approaching the History of the Fossil Economy." *Critical Historical Studies* 3, no. 2. https://doi.org/10.1086/688347.

____. 2021a. *How to Blow Up a Pipeline*. London: Verso. EPUB.

____. 2021b. "No Safe Options: A Conversation with Andreas Malm." Interview by Wen Stephenson. *LA Review of Books*, January 5, 2021. lareviewofbooks.org/article/no-safe-options-a-conversation-with-andreas-malm.

____. 2021c. "Planning the Planet: Geoengineering Our Way Out of and Back into a Planned Economy." In *Has It Come to This? The Promises and Perils of Geoengineering on the Brink*, edited by J.P. Sapinsky, Holly Jean Buck, and Andreas Malm. New Brunswick, NJ: Rutgers University Press.

Malm, Andreas, and Wim Carton. 2021. "Seize the Means of Carbon Removal: The Political Economy of Direct Air Capture." *Historical Materialism* 29, no. 1. https://doi.org/10.1163/1569206X-29012021.

Malm, Andreas, and the Zetkin Collective. 2021. *White Skin, Back Fuel: On the Danger of Fossil Fascism.* London: Verso.

Mangat, Rupinder, Simon Dalby, and Matthew Paterson. 2018. "Divestment Discourse: War, Justice, Morality and Money." *Environmental Politics* 27, no. 2. https://doi.org/10.1080/09644016.2017.1413725.

Mann, Michael. 2021. *The New Climate War: The Fight to Take Back the Planet.* New York: Public Affairs.

Marois, Thomas, and Ali Rıza Güngen. 2019. *A US Green Investment Bank for All: Democratized Finance for a Just Transition.* Washington, DC: The Next System Project.

Mastini, Riccardo, Giorgos Kallis, and Jason Hickel. 2021. "A Green New Deal without Growth?" *Ecological Economics* 179. https://doi.org/10.1016/j.ecolecon.2020.106832.

McBay, Aric, Lierre Keith, and Derrick Jensen. 2011. *Deep Green Resistance: Strategy to Save the Planet.* New York: Seven Stories Press.

McCarthy, Tom, and Alvin Chang. 2021. " 'The Senate Is Broken': System Empowers White Conservatives, Threatening US Democracy." *The Guardian*, March 13, 2021. theguardian.com/us-news/2021/mar/12/us-senate-system-white-conservative-minority.

McConnell, Mitch. 2019a. "Green New Deal: Pain for American Families, No Meaningful Change in Carbon Emissions." *Republican Leader*, March 26, 2019. republicanleader.senate.gov/newsroom/remarks/green-new-deal-pain-for-american-families-no-meaningful-change-in-carbon-emissions.

___. 2019b. "Green New Deal: Immediate Harm for American Workers and Families." *Republican Leader*, March 13, 2019. republicanleader.senate.gov/newsroom/remarks/green-new-deal-immediate-harm-for-american-workers-and-families.

McCright, Aaron M. 2016. "Anti-Reflexivity and Climate Change Skepticism in the US General Public." *Human Ecology Review* 22, no. 2. https://doi.org/10.4225/13/58213a5387787.

McCright, Aaron M., and Riley E. Dunlap. 2010. "Anti-Reflexivity: The American Conservative Movement's Success in Undermining Climate Science and Policy." *Theory, Culture & Society* 27, no. 2–3. https://doi.org/10.1177/0263276409356001.

McGraw, Seamus. 2018. "Necessity Defense Goes Untested in Climate Activists' Acquittal." *The Climate Docket*, October 9, 2018. climatedocket.com/2018/10/09/necessity-defense-climate-change-trial-2.

McKibben, Bill. 2010. *Earth: Making a Life on a Tough New Planet.* New York: Henry Holt and Company.

___. 2012. "Global Warming's Terrifying New Math," *Rolling Stone*, July 19, 2012. rollingstone.com/politics/politics-news/global-warmings-terrifying-new-math-188550.

___. 2016. "A World at War." *The New Republic*, August 15, 2016. newrepublic. com/article/135684/declare-war-climate-change-mobilize-wwii.

___. 2019a. *Falter: Has the Human Game Begun to Play Itself Out?* New York: Henry Holt and Company.

___. 2019b. "The Hard Lessons of Dianne Feinstein's Encounter with the Young Green New Deal Activists." *The New Yorker*, February 23, 2019. newyorker. com/news/daily-comment/the-hard-lessons-of-dianne-feinsteins-encounter-with-the-young-green-new-deal-activists-video.

___. 2019c. "The New Climate Math: The Numbers Keep Getting More Frightening." *Yale Environment 360*, November 25, 2019. e360.yale.edu/features/the-new-climate-math-the-numbers-keep-getting-more-frightening.

McKinnon, Catriona. 2019. "The Panglossian Politics of the Geoclique." *Critical Review of International Social and Political Philosophy* 23, no. 5. https://doi. org/10.1080/13698230.2020.1694216.

Meade, Amanda. 2020. "The Australian: Murdoch-Owned Newspaper Accused of Downplaying Bushfires in Favour of Picnic Races." *The Guardian*, January 3, 2020. theguardian.com/media/2020/jan/04/ the-australian-murdoch-owned-newspaper-accused-of-downplaying-bushfires-in-favour-of-picnic-races.

Meadows, Donella H., Dennis L. Meadows, Jørgen Randers, and William W. Behrens III. 1972. *The Limits to Growth*. New York: Universe Books.

Media Matters Staff. 2019. "Here Are Some of the Dumbest Right-Wing Media Takes on the Green New Deal." February 8, 2019. mediamatters. org/sean-hannity/here-are-some-dumbest-right-wing-media-takes-green-new-deal.

Meiksins Wood, Ellen. 1995. *Democracy Against Capitalism: Renewing Historical Materialism*. Cambridge, UK: Cambridge University Press.

Meyer, Thomas, with Lewis Hinchman. 2007. *The Theory of Social Democracy*. Malden, MA: Polity.

Michelin, Ossie. 2020. " 'Solastalgia': Arctic Inhabitants Overwhelmed by New Form of Climate Grief." *The Guardian*, October 15, 2020. theguardian.com/ us-news/2020/oct/15/arctic-solastalgia-climate-crisis-inuit-indigenous.

Ministry of Just Transition Collective. 2022. "The Year Is 2025, and a Just Transition Has Transformed Canada." *The Breach*, March 18, 2022. breach-media.ca/the-year-is-2025-and-a-just-transition-has-transformed-canada%ef%bf%bc.

Mill, John Stuart. [1848] 1988. *Principles of Political Economy, Books IV and V*. Penguin.

Millhiser, Ian. 2021. "Build Back Better Is the Latest Victim of America's Anti-Democratic Senate." *Vox*, December 20, 2021. vox. com/2021/12/20/22846504/senate-joe-manchin-build-back-better-democrats-republicans-43-million.

Millward-Hopkins, Joel, Julia K. Steinberger, Narasimha D. Rao, and Yannick

Oswald. 2020. "Providing Decent Living with Minimum Energy: A Global Scenario." *Global Environmental Change* 65. https://doi.org/10.1016/j.gloenvcha.2020.102168.

Milman, Oliver. 2021. "Joe Manchin Leads Opposition to Biden's Climate Bill, Backed by Support from Oil, Gas and Coal." *The Guardian*, October 20, 2021. theguardian.com/us-news/2021/oct/20/joe-manchin-oil-and-gas-fossil-fuels-senator.

Moellendorf, Darrell. 2014. *The Moral Challenge of Dangerous Climate Change: Values, Poverty, and Policy*. New York: Cambridge University Press.

Monbiot, George. 2015. "There May Be Flowing Water on Mars. But Is There Intelligent Life on Earth?" *The Guardian*, September 29, 2015. theguardian.com/commentisfree/2015/sep/29/water-mars-intelligent-life-earth-nasa.

___. 2017a. "How Labour Could Lead the Global Economy Out of the 20th Century." *The Guardian*, October 11, 2017. theguardian.com/commentisfree/2017/oct/11/labour-global-economy-planet.

___. 2017b. *Out of the Wreckage: A New Politics for an Age of Crisis*. London: Verso. EPUB.

___. 2019. "Today. I Aim to Get Arrested. It Is the Only Real Power Climate Protesters Have." *The Guardian*, October 16, 2019. theguardian.com/commentisfree/2019/oct/16/i-aim-to-get-arrested-climate-protesters.

Moore, Jason. 2015. *Capitalism in the Web of Life: Ecology and the Accumulation of Capital*. Brooklyn, NY: Verso.

Morton, Adam, and Ben Smee. 2019. "Great Barrier Reef Expert Panel Says Peter Ridd Misrepresenting Science." *The Guardian*, August 27, 2019. theguardian.com/environment/2019/aug/28/great-barrier-reef-expert-panel-says-peter-ridd-misrepresenting-science.

Morton, Oliver. 2015. *The Planet Remade: How Geoengineering Could Change the World*. Princeton, NJ: Princeton University Press.

Moser, Claire, and Ryan Koronowski. 2017. "The Climate Denier Caucus in Trump's Washington," *ThinkProgress*, April 28, 2017. archive.thinkprogress.org/115th-congress-climate-denier-caucus-65fb825b3963/.

Mueller-Hsia, Kaylana. 2021. "Anti-Protest Laws Threaten Indigenous and Climate Movements." *Brennan Center for Justice*, March 17, 2021. brennancenter.org/our-work/analysis-opinion/anti-protest-laws-threaten-indigenous-and-climate-movements.

Muraca, Barbara. 2013. "Décroissance: A Project for a Radical Transformation of Society." *Environmental Values* 22. https://doi.org/10.3197/096327113X13581561725112.

Nakate, Vanessa (@vanessa_vash). 2020. "Share if you can […] What it means to be removed from a photo!" *Twitter*, January 24, 2020, 7:57 a.m. twitter.com/vanessa_vash/status/1220722317002556098.

Neale, Johnathan. 2021. "Lithium, Batteries and Climate Change." *Climate and Capitalism*, February 11, 2021. climateandcapitalism.com/2021/02/11/lithium-batteries-and-climate-change.

Nixon, Rob. 2011. *Slow Violence and the Environmentalism of the Poor.* Cambridge, MA: Harvard University Press.

Normann, Susanne. 2021. "Green Colonialism in the Nordic Context: Exploring Southern Saami Representations of Wind Energy Development." *Journal of Community Psychology* 49, no. 1. https://doi.org/10.1002/jcop.22422.

Nussbaum, Martha C. 2013. *Creating Capabilities: The Human Development Approach.* Cambridge, MA: Belknap.

O'Neill, Daniel W. 2018. "Is It Possible for Everyone to Live a Good Life Within Our Planet's Limits?" *The Conversation*, February 7, 2018. theconversation.com/is-it-possible-for-everyone-to-live-a-good-life-within-our-planets-limits-91421.

O'Neill, Daniel W., Andrew L. Fanning, William F. Lamb, and Julia K. Steinberger. 2018. "A Good Life for All Within Planetary Boundaries." *Nature Sustainability* 1. https://doi.org/10.1038/s41893-018-0021-4.

Office of the Parliamentary Budget Officer. 2020. *Estimating the Top Tail of the Family Wealth Distribution in Canada.* pbo-dpb.gc.ca/web/default/files/Documents/Reports/RP-2021-007-S/RP-2021-007-S_en.pdf.

Oreskes, Naomi, and Erik M. Conway. 2010. *Merchants of Doubt: How a Handful of Scientists Obscured the Truth on Issues from Tobacco Smoke to Global Warming.* New York: Bloomsbury.

Ott, Konrad. 2012. "Variants of De-Growth and Deliberative Democracy: A Habermasian Proposal." *Futures* 44 no. 6. https://doi.org/10.1016/j.futures.2012.03.018.

Oxfam. 2015. "Extreme Carbon Inequality: Why the Paris Climate Deal Must Put the Poorest, Lowest Emitting and Most Vulnerable People First." hdl.handle.net/10546/582545.

____. 2020a. *Climate Finance Shadow Report 2020: Assessing Progress towards the $100 Billion Commitment.* hdl.handle.net/10546/621066.

____. 2020b. *Confronting Carbon Inequality.* oxfam.org/en/research/extreme-carbon-inequality.

Page, Edward A. 2006. *Climate Change, Justice and Future Generations.* Cheltenham, UK: Edward Elgar Publishing.

Pape, Robert A. 2022. "The Jan. 6 Insurrectionists Aren't Who You Think They Are." *Foreign Policy*, January 6, 2022. foreignpolicy.com/2022/01/06/trump-capitol-insurrection-january-6-insurrectionists-great-replacement-white-nationalism.

Parenti, Christian. 2011. *Tropic of Chaos: Climate Change and the New Geography of Violence.* New York: Nation Books.

____. 2021. "A Left Defense of Carbon Dioxide Removal." In *Has It Come to This? The Promises and Perils of Geoengineering on the Brink*, edited by J.P. Sapinsky, Holly Jean Buck, and Andreas Malm. New Brunswick, NJ: Rutgers University Press.

Payal, Parekh (@payalclimate). 2019. "Msg I received on FB." *Twitter*, April 16, 2019, 9:11 a.m. twitter.com/payalclimate/status/1118170044096184322.

People's Climate March. 2014. "Frequently Asked Questions." docs.google.com/document/d/1PXv54n8r6kkO0A2ijynwGAQcAVoNvsGIxzdNjubr_0o/edit.

Perry, Elizabeth. 2020. "Linking the Crises of Covid-19, Environmental Justice, and Police Violence – Updated." *Work and Climate Change Report*, June 5, 2020. workandclimatechangereport.org/2020/06/05/linking-the-crises-of-covid-19-racism-environmental-justice-and-police-violence.

Perkins, Patricia E. (Ellie). 2019. "Climate Justice, Commons, and Degrowth." *Ecological Economics* 160. https://doi.org/10.1016/j.ecolecon.2019.02.005.

Pesca, Mike. 2019. "The Green New Deal Will Never Work." *Slate*, February 8, 2019. slate.com/news-and-politics/2019/02/green-new-deal-unrealistic-impossible-experts.html.

Phillips, Leigh. 2015. *Austerity Ecology & the Collapse-Porn Addicts: A Defence of Growth, Progress, Industry and Stuff*. Winchester, UK: Zero Books.

Pielke, Roger A., Jr. 2010. "A Positive Path for Meeting the Global Climate Challenge." *Yale Environment 360*, October 18, 2010. e360.yale.edu/features/a_positive_path_for_meeting_the_global_climate_challenge.

Pinker, Steven. 2018. *Enlightenment Now: The Case for Reason, Science, Humanism, and Progress*. New York: Viking. EPUB.

Pollin, Robert. 2018. "Degrowth vs. a Green New Deal." *New Left Review* 112. newleftreview.org/issues/ii112/articles/robert-pollin-de-growth-vs-a-green-new-deal.

____. 2019. "Degrowth versus Green New Deal: Response to Juliet Schor and Andrew Jorgenson." *Review of Radical Political Economics* 51, no. 2. https://doi.org/10.1177/0486613419833522.

____. 2020a. "An Industrial Policy Framework to Advance a Global Green New Deal." In *The Oxford Handbook of Industrial Policy*, edited by Arkebe Oqubay, Christopher Cramer, Ha-Joon Chang, and Richard Kozul-Wright. Oxford, UK: Oxford University Press. https://doi.org/10.1093/oxfordhb/9780198862420.013.16.

____. 2020b. "Noam Chomsky's Green New Deal." Interview by David Roberts. *Vox*, September 21, 2020. vox.com/energy-and-environment/21446383/noam-chomsky-robert-pollin-climate-change-book-green-new-deal.

____. 2022. "Nationalize the U.S. Fossil Fuel Industry to Save the Planet." *American Prospect*, April 8, 2022. prospect.org/environment/nationalize-us-fossil-fuel-industry-to-save-the-planet.

Pollin, Robert, and Brian Callaci. 2019. "The Economics of Just Transition: A Framework for Supporting Fossil Fuel–Dependent Workers and Communities in the United States." *Labor Studies Journal* 44, no. 2. https://doi.org/10.1177/0160449X18787051.

Price, Carter C., and Kathryn A. Edwards. 2020. *Trends in Income From 1975 to 2018*. Santa Monica, CA: RAND Education and Labor. rand.org/pubs/working_papers/WRA516-1.html.

Public Citizen. 2019. *Foxic: Fox News Network's Dangerous Climate Denial*

2019. citizen.org/article/foxic-fox-news-networks-dangerous-climate-denial-2019.

Quilley, Stephen. 2013. "De-Growth Is Not a Liberal Agenda: Relocalisation and the Limits to Low Energy Cosmopolitanism." *Environmental Values* 22: 261–285. https://doi.org/10.3197/096327113X13581561725310.

Rahman, Minnie. 2019. "This Is What Extinction Rebellion Must Do to Engage with People of Colour on Climate Justice." *gal-dem*, April 30, 2019. gal-dem.com/this-is-what-extinction-rebellion-must-do-to-engage-with-people-of-colour-on-climate-justice.

Rakia, Raven. 2015. "Ranchers with Ties to the Biofuel Industry Attack Brazilian Tribe Members." *Grist*, September 24, 2015. grist.org/article/ranchers-with-ties-to-the-biofuel-industry-attack-brazilian-tribe-members/.

Ramírez, Nikki McCann. 2022. "Tucker Carlson's History of Fearmongering about White Replacement, Genocide, and Race War." *Media Matters*, July 1, 2021 (updated May 14, 2022). mediamatters.org/tucker-carlson/tucker-carlsons-history-fearmongering-about-white-replacement-genocide-and-race-war.

Rand, Tom. 2020. *The Case for Climate Capitalism: Economic Solutions for a Planet in Crisis*. Toronto: ECW Press. EPUB.

Raworth, Kate. 2012a. "The Economic Vandal Strikes Back." *Kate Raworth* (blog), August 3, 2012. kateraworth.com/2012/08/03/the-economic-vandal-strikes-back.

___. 2012b. "Why It's Time to Vandalize the Economic Textbooks." *Kate Raworth* (blog), July 23, 2012. kateraworth.com/2012/07/23/why-its-time-to-vandalize-the-economic-textbooks.

___. 2017. *Doughnut Economics: 7 Ways to Think Like a 21st Century Economist*. White River Junction, VT: Chelsea Green.

Readfearn, Graham. 2019. "Australia's Science Academy Attacks 'Cherrypicking' of Great Barrier Reef Research." *The Guardian*, November 26, 2019. theguardian.com/environment/2019/nov/26/australias-science-academy-attacks-cherrypicking-of-great-barrier-reef-research.

Rehman, Asad. 2019. "The 'Green New Deal' Supported by Ocasio-Cortez and Corbyn Is Just a New Form of Colonialism." *Independent*, May 4, 2019. independent.co.uk/voices/green-new-deal-alexandria-ocasio-cortez-corbyn-colonialism-climate-change-a8899876.html.

Ritchie, Hannah. 2019. "Who Has Contributed Most to Global Co2 Emissions?" *Our World in Data*, October 1, 2019. ourworldindata.org/contributed-most-global-co2.

Roberts, David. 2016. "Is It Useful to Think of Climate Change as a 'World War'?" *Vox*, August 18, 2016. vox.com/2016/8/18/12507810/climate-change-world-war.

___. 2018a. "Washington Votes No on a Carbon Tax — Again." *Vox*, September 28, 2018. vox.com/energy-and-environment/2018/9/28/17899804/washington-1631-results-carbon-fee-green-new-deal.

———. 2018b. "We Now Have a Dollar Value for One of Oil's Biggest Subsidies." *Vox*, September 21, 2018. vox.com/energy-and-environment/2018/9/21/17885832/oil-subsidies-military-protection-supplies-safe.

———. 2020a. "At Last, a Climate Policy Platform that Can Unite the Left." *Vox*, July 9, 2020. vox.com/energy-and-environment/21252892/climate-change-democrats-joe-biden-renewable-energy-unions-environmental-justice.

———. 2020b. "How to Build a Circular Economy that Recycles Carbon." *Vox*, January 8, 2020. vox.com/energy-and-environment/2020/1/8/20841897/climate-change-carbon-capture-circular-economy-recycle.

———. 2021a. "Can the US Reach Biden's Climate Goal without the CEPP?" *Volts*, October 20, 2021. volts.wtf/p/can-the-us-reach-bidens-climate-goal.

———. 2021b. "There Is No "Moderate" Position on Climate Change." *Volts*, June 30, 2021. volts.wtf/p/there-is-no-moderate-position-on?s=r.

———. 2022a. "Minerals and the Clean-Energy Transition: The Basics." *Volts*, January 21, 2022. volts.wtf/p/minerals-and-the-clean-energy-transition.

———. 2022b. "The Minerals Used by Clean-Energy Technologies." *Volts*, February 7, 2022. volts.wtf/p/the-minerals-used-by-clean-energy.

Roberts, J. Timmons, and Bradley C. Parks. 2007. *A Climate of Injustice: Global Inequality, North-South Politics, and Climate Policy.* Cambridge, MA: MIT Press.

Robinson, Kim Stanley. 2018. "The King of Climate Fiction Makes the Left's Case for Geoengineering." Interview by Alexander C. Kaufman. *HuffPost*, July 28, 2018. huffpost.com/entry/climate-geoengineering-kim-stanley-robinson_n_5b4e54bde4b0de86f487b0b9.

Robinson, Nathan J. 2020. "The Last-Ditch Talking Point on Climate Change." *Current Affairs*, September 15, 2020. currentaffairs.org/2020/09/the-last-ditch-talking-point-on-climate-change.

Rockström, Johan, Will Steffen, Kevin Noone, et al. 2009. "A Safe Operating Space for Humanity." *Nature* 461. https://doi.org/10.1038/461472a.

Romm, Joe. 2017. "Trump's Reported Exit from Paris Climate Deal Signals End of the American Century." *Think Progress.* May 31, 2017. archive.thinkprogress.org/trump-paris-end-of-the-american-century-ec5ee0742f8a/.

Rowell, Andy. 2020. "Denial to the Death: In Australia, Newspaper Headlines Tout 'Warming Is Good for Us.' " *Oil Change International*, January 27, 2020. priceofoil.org/2020/01/27/denial-to-the-death-in-australia-newspaper-headlines-tout-warming-is-good-for-us.

Rupar, Aaron. 2019. "CPAC Speakers Keep Saying Democrats Want to Ban Cows and Legalize Infanticide. They Don't." *Vox*, March 2, 2019. vox.com/2019/3/2/18246812/cpac-2019-themes-cows-infanticide-don-jr-pence-meadows.

Saad, Aaron. 2017. "Toward a Justice Framework for Understanding and Responding to Climate Migration and Displacement." *Environmental Justice*

10, no. 4. https://doi.org/10.1089/env.2016.0033.

___. 2019a. "Conspiracy Theory of 'Foreign-Funded' Tar Sands Opposition Reveals Ugly Truth." *Ricochet*, August 14, 2019. ricochet.media/en/2699/conspiracy-theory-of-foreign-funded-tar-sands-opposition.

___. 2019b. "What's Necessary for Good People to Do Nothing." *Ricochet*, September 5, 2019. ricochet.media/en/2712/whats-necessary-for-good-people-to-do-nothing.

Saez, Emmanuel, and Gabriel Zucman. 2019a. "Progressive Wealth Taxation." *Brookings Papers on Economic Activity*, Fall. brookings.edu/bpea-articles/progressive-wealth-taxation.

___. 2019b. *The Triumph of Injustice: How the Rich Dodge Taxes and How to Make Them Pay.* New York: W.W. Norton & Company.

Samios, Zoe, and Andrew Hornery. 2020. "'Dangerous, Misinformation': News Corp Employee's Fire Coverage Email." *Sydney Morning Herald*, January 10, 2020. smh.com.au/environment/climate-change/dangerous-misinformation-news-corp-employee-s-fire-coverage-email-20200110-p53qel.html.

Savard, Alain. 2019. "How Seven Thousand Quebec Workers Went on Strike against Climate Change." *Labor Notes*, October 25, 2019. labornotes.org/2019/10/how-seven-thousand-quebec-workers-went-strike-against-climate-change.

Schapper, Andrea. 2018. "Climate Justice and Human Rights." *International Relations* 32, no. 3. https://doi.org/10.1177/0047117818782595.

Schlosser, Kolson. 2020. "Contrasting Visions of the Green New Deal." *Environmental Politics* 30, no. 3. https://doi.org/10.1080/09644016.2020.1847514.

Schneider, Linda, and Lili Fuhr. 2020. "Defending a Failed Status Quo: The Case against Geoengineering from a Civil Society Perspective." In *Has It Come to This? The Promises and Perils of Geoengineering on the Brink*, edited by J.P. Sapinsky, Holly Jean Buck, and Andreas Malm. New Brunswick, NJ: Rutgers University Press.

Schwartz, John. 2017. " 'A Conservative Climate Solution': Republican Group Calls for Carbon Tax." *New York Times*, February 7, 2017. nytimes.com/2017/02/07/science/a-conservative-climate-solution-republican-group-calls-for-carbon-tax.html.

Scranton, Roy. 2015. *Learning to Die in the Anthropocene: Reflections on the End of Civilization.* San Francisco: City Lights Books.

Serwer, Adam. 2018. "The Cruelty Is the Point." *Atlantic*, October 3, 2018. theatlantic.com/ideas/archive/2018/10/the-cruelty-is-the-point/572104.

Shao, Qing-long, and Beatriz Rodríguez-Labajos. 2016. "Does Decreasing Working Time Reduce Environmental Pressures? New Evidence Based on Dynamic Panel Approach." *Journal of Cleaner Production* 125. https://doi.org/10.1016/j.jclepro.2016.03.037.

Shapiro, Ben. 2019. "AOC's Green New Deal Proposal Is One of The Stupidest Documents Ever Written." *Daily Wire*, February 7, 2019. dailywire.com/

news/aocs-green-new-deal-proposal-one-stupidest-ben-shapiro.

Shiva, Vandana. 2008. *Soil Not Oil: Environmental Justice in an Age of Climate Crisis*. Boston: South End Press.

Shulman, Seth, Kate Abend, and Alden Meyer. 2007. *Smoke, Mirrors & Hot Air: How ExxonMobil Uses Big Tobacco's Tactics to Manufacture Uncertainty on Climate Science*. Cambridge, MA: Union of Concerned Scientists.

Silk, Ezra. 2019. *Victory Plan*. Version revised by Kaela Bamberger. The Climate Mobilization. theclimatemobilization.org/resources/whitepapers/victory-plan.

Simms, Andrew. 2009. *Ecological Debt: Global Warming and the Wealth of Nations*. London: Pluto Press.

Skidelsky, Robert, and Ed Skidelsky. 2013. *How Much Is Enough? Money and the Good Life*. London: Penguin.

Smith, Noah. 2019. "The Green New Deal Would Spend the U.S. into Oblivion." *Bloomberg*, February 8, 2019. bloomberg.com/opinion/articles/2019-02-08/alexandria-ocasio-cortez-s-green-new-deal-is-unaffordable.

___. 2021. "People Are Realizing that Degrowth Is Bad." *Noahpinion*, September 6, 2021. noahpinion.substack.com/p/people-are-realizing-that-degrowth.

Snyder, Timothy. 2017. *On Tyranny: Twenty Lessons from the Twentieth Century*. New York: Tim Duggan Books.

___. 2021. "The American Abyss." *New York Times*, January 9, 2021. nytimes.com/2021/01/09/magazine/trump-coup.html.

Socialist Resistance. 2020. "Ecosocialism: The Strategic Debate." October 15, 2020, socialistresistance.org/ecosocialism-the-strategic-debate/21010.

Solon, Pablo. 2014. "How Did Leaders Respond to the People's Climate March?" *teleSUR*, September 27, 2014. telesurtv.net/opinion/How-Did-Leaders-Respond-to-the-Peoples-Climate-March-20140927-0058.html.

Solnit, Rebecca. 2014. "Call Climate Change What It Is: Violence." *The Guardian*, April 7, 2014. theguardian.com/commentisfree/2014/apr/07/climate-change-violence-occupy-earth.

Soper, Kate. 2020. *Post-Growth Living: For an Alternative Hedonism*. London: Verso.

Spring, Jake. 2021. "Geoengineering Marks Scientific Gains in U.N. Report on Dire Climate Future." *Reuters*, August 10, 2021. reuters.com/business/environment/geoengineering-marks-scientific-gains-un-report-dire-climate-future-2021-08-10.

Spross, Jeff, Tiffany Germain, and Ryan Koronowski. 2013. "The Anti-Science Climate Denier Caucus: 113th Congress Edition." *ThinkProgress*, June 26, 2013. archive.thinkprogress.org/the-anti-science-climate-denier-caucus-113th-congress-edition-82ef03690c02/.

Srnicek, Nick, and Alex Williams. 2015. *Inventing the Future: Postcapitalism and a World Without Work*. London: Verso.

Standing, Guy. 2011. *The Precariat: The New Dangerous Class*. London: Bloomsbury.

___. 2017. *Basic Income: And How We Can Make It Happen*. London, UK: Pelican Books.

Steffen, Will, Katherine Richardson, Johan Rockström, et al. 2015. "Planetary Boundaries: Guiding Human Development on a Changing Planet." *Science* 347, no. 6223. https://doi.org/10.1126/science.1259855.

Stern, Nicholas. 2007. *The Economics of Climate Change: The Stern Review*. Cambridge, UK: Cambridge University Press.

Stop Funding Heat. 2021. *#InDenial – Facebook's Growing Friendship with Climate Misinformation*. stopfundingheat.info/facebook-in-denial.

Stuart, Diana, Ryan Gunderson, and Brian Petersen. 2020. "The Climate Crisis as a Catalyst for Emancipatory Transformation: An Examination of the Possible." *International Sociology* 35, no. 4. https://doi.org/10.1177/0268580920915067.

Supran, Geoffrey, and Naomi Oreskes. 2020. "ExxonMobil Misled the Public about the Climate Crisis. Now They're Trying to Silence Critics." *The Guardian*, October 16, 2020. theguardian.com/commentisfree/2020/oct/16/exxonmobil-misled-the-public-about-the-climate-crisis-now-theyre-trying-to-silence-critics.

Swaminathan, Nikhil. 2019. "Tauntauns, Seahorses, and Lotsa Babies: Mike Lee Trolls the Green New Deal." *Grist*, March 26, 2019. grist.org/article/senator-mike-lee-green-new-deal-climate-change.

Táíwò, Olúfẹ́mi O. 2020. "Climate Apartheid Is the Coming Police Violence Crisis." *Dissent*, August 12, 2020. dissentmagazine.org/online_articles/climate-apartheid-is-the-coming-police-violence-crisis.

Theel, Shauna, Jill Fitzsimmons, and Max Greenberg. 2012. "TIMELINE: Fox News' Role in the 'Climate of Doubt.' " *Media Matters*, October 24, 2012. mediamatters.org/sean-hannity/timeline-fox-news-role-climate-doubt.

Thomas, Kimberley, R. Dean Hardy, Heather Lazrus, et al. 2019. "Explaining Differential Vulnerability to Climate Change: A Social Science Review." *WIRES Climate Change* e565. https://doi.org/10.1002/wcc.565.

Treen, Kathie M. d'I., Hywel T. P. Williams, and Saffron J. O'Neill. 2020. "Online Misinformation about Climate Change." *WIREs Climate Change* 11, no. 5. https://doi.org/10.1002/wcc.665.

UNCTAD (United Nations Conference on Trade and Development). 2019. *Trade and Development Report 2019: Financing a Global Green New Deal*. Geneva: United Nations.

United Nations Development Programme. 2007. *Human Development Report 2007/8; Fighting Climate Change: Human Solidarity in a Divided World*. New York: UNDP. hdr.undp.org/en/content/human-development-report-20078.

United Nations Environment Programme and Sabin Center for Climate Change Law. 2020. *Global Climate Litigation Report: 2020 Status Review*. Nairobi: UNEP. unep.org/resources/report/global-climate-litigation-report-2020-status-review.

Union of Concerned Scientists. 2017. "How Fossil Fuel Lobbyists Used 'Astroturf' Front Groups to Confuse the Public." October 17, 2017. ucsusa.

org/resources/how-fossil-fuel-lobbyists-used-astroturf-front-groups-confuse-public.

Victor, Peter A. 2008. *Managing Without Growth: Slower by Design, Not Disaster*. Northampton, MA: Edward Elgar.

___. 2012. "Growth, Degrowth and Climate Change: A Scenario Analysis." *Ecological Economics* 84. https://doi.org/10.1016/j.ecolecon.2011.04.013.

Vidal, John. 2008. "Not Guilty: The Greenpeace Activists Who Used Climate Change as a Legal Defence." *The Guardian*, September 11, 2008. theguardian.com/environment/2008/sep/11/activists.kingsnorthclimatecamp.

Wahlström, Mattias, Magnus Wennerhag, and Christopher Rootes. 2013. "Framing 'The Climate Issue': Patterns of Participation and Prognostic Frames among Climate Summit Protesters." *Global Environmental Politics* 13, no. 4. https://doi.org/10.1162/GLEP_a_00200.

Wainwright, Joel, and Geoff Mann. 2018. *Climate Leviathan: A Political Theory of Our Planetary Future*. London: Verso.

Waldman, Paul. 2020. "Hatred of Liberals Is All that's Left of Conservatism." *Washington Post*, December 11, 2020. washingtonpost.com/opinions/2020/12/11/hatred-liberals-is-all-thats-left-conservatism.

Waldman, Scott. 2019. "Group Tied to Shadowy Network Created $93T Estimate." *E&E News*, April 1, 2019. eenews.net/stories/1060137815.

Waldman, Scott, and Benjamin Hulac. 2018. "This Is When the GOP Turned Away from Climate Policy." *E&E News*, December 5, 2018. eenews.net/stories/1060108785.

Walton, Kate. 2020. "How 'Murdochracy' Controls the Climate Debate in Australia." *Al-Jazeera*, January 24, 2020. aljazeera.com/news/2020/1/24/how-murdochracy-controls-the-climate-debate-in-australia.

Warlenius, Rikard. 2018. "Decolonizing the Atmosphere: The Climate Justice Movement on Climate Debt." *Journal of Environment & Development* 27, no. 2. https://doi.org/10.1177/1070496517744593.

We Mean Business Coalition. 2021. "More Than 100 Multinational Corporations Have Taken the Climate Pledge." *Climate Home News*, April 23, 2021. climatechangenews.com/2021/04/23/100-multinational-corporations-taken-climate-pledge.

Whyte, Kyle Powys. 2017. "Is It Colonial Deja Vu? Indigenous Peoples and Climate Injustice." In *Humanities for the Environment: Integrating Knowledges, Forging New Constellations of Practice*, edited by J. Adamson, M. Davis, and H. Huang. London: Routledge.

___. 2018. "Indigeneity in Geoengineering Discourses: Some Considerations." *Ethics, Policy & Environment* 21, no. 3. https://doi.org/10.1080/21550085.2018.1562529.

___. 2019. "Way Beyond the Lifeboat: An Indigenous Allegory of Climate Justice." In *Climate Futures: Reimagining Global Climate Justice*, edited by Debashish Munshi, Kum-Kum Bhavnani, John Foran, and Priya Kurian. London: Zed Books.

___. 2021. "Geoengineering and Indigenous Climate Justice: A Conversation with Kyle Powys Whyte." Interview by Holly Jean Buck. In *Has It Come to This? The Promises and Perils of Geoengineering on the Brink*, edited by J.P. Sapinsky, Holly Jean Buck, and Andreas Malm. New Brunswick, NJ: Rutgers University Press.

Wiedmann, Thomas, Manfred Lenzen, Lorenz T. Keyßer, and Julia K. Steinberger. 2020. "Scientists' Warning on Affluence." *Nature Communications* 11. https://doi.org/10.1038/s41467-020-16941-y.

Wilkinson, Richard G., and Kate Pickett. 2011. *The Spirit Level: Why Greater Equality Makes Societies Stronger*. New York: Bloomsbury.

Wilson, Cameron. 2020. "Climate Deniers Are Making Memes about the Coronavirus to Argue against Urgent Climate Action." *Buzzfeed*, April 1, 2020. buzzfeed.com/cameronwilson/right-wing-coronavirus-climate-change-memes-denier.

Wilt, James. 2021. "How to Blow Up a Movement: Andreas Malm's New Book Dreams of Sabotage but Ignores Consequences." *Canadian Dimension*, March 3, 2021. canadiandimension.com/articles/view/how-to-blow-up-a-movement-malms-new-book-dreams-of-sabotage-but-ignores-consequences.

World Bank. 2020. *Minerals for Climate Action: The Mineral Intensity of the Clean Energy Transition*. Washington, DC: World Bank. pubdocs.worldbank.org/en/961711588875536384/Minerals-for-Climate-Action-The-Mineral-Intensity-of-the-Clean-Energy-Transition.pdf.

___. 2021. *State and Trends of Carbon Pricing 2021*. World Bank, Washington, DC. https://doi.org/10.1596/978-1-4648-1728-1.

___. 2022. *State and Trends of Carbon Pricing 2022*. Washington, DC: World Bank. hdl.handle.net/10986/37455.

World People's Conference on Climate Change and the Rights of Mother Earth. 2010a. "Working Group 8: Climate Debt." pwccc.wordpress.com/2010/04/16/working-group-8-climate-debt.

___. 2010b. "Final Conclusions Working Group 2: Harmony with Nature to Live Well." pwccc.wordpress.com/2010/04/30/final-conclusions-working-group-2-harmony-with-nature-to-live-well.

Wretched of the Earth. 2015. "Indigenous People Were Silenced and Erased." *New Internationalist*, December 17, 2015. newint.org/blog/2015/12/17/wretched-of-the-earth-open-letter.

___. 2019. "An Open Letter to Extinction Rebellion." *Society for the Diffusion of Useful Knowledge* 6. blackwoodgallery.ca/publications/sduk/forging/an-open-letter-to-extinction-rebellion.

Wright, Erik Olin. 2010. *Envisioning Real Utopias*. London: Verso.

___. 2019. *How to Be an Anticapitalist in the Twenty-First Century*. London: Verso. EPUB.

WWF (World Wide Fund for Nature). 2020. *Living Planet Report 2020: Bending the Curve of Biodiversity Loss*. Gland, Switzerland: WWF.

WWF-Australia (World Wide Fund for Nature Australia). 2020. *Australia's 2019–2020 Bushfires: The Wildlife Toll. Interim Report.* wwf.org.au/ArticleDocuments/353/Animals%20Impacted%20Interim%20Report%20 24072020%20final.pdf.aspx.

Yglesias, Matthew. 2019. "Alexandria Ocasio-Cortez Is Floating a 70 Percent Top Tax Rate — Here's the Research that Backs Her Up." *Vox*, January 7, 2019. vox.com/policy-and-politics/2019/1/4/18168431/alex-andria-ocasio-cortez-70-percent.

Yourish, Karen, Weiyi Cai, Larry Buchanan, et al. 2022. "Inside the Apocalyptic Worldview of 'Tucker Carlson Tonight.' " *New York Times*, April 30, 2022. nytimes.com/interactive/2022/04/30/us/tucker-carlson-tonight.html.

INDEX

Monbiot, George, 131, 144
motivated reasoning, 21
Musk, Elon, 96

Nakate, Vanessa, 179, 180
National Review, 97
Necessity defence, 178–179
negative externality, 51, 52, 106, 116,
    178, 191
neoliberalism
    climate solutions and, 51–58
    critiques of, 50, 113, 116–118
    explanations for what has pre-
        vented climate policy that are
        consistent with, 61
    geoengineering and, 97, 99, 103,
        104, 106, 107
    main beliefs of, 47–50
    obstacles to climate action created
        by, 39, 43, 47, 48, 63–66, 72,
        116
New Deal, 115
News Corp, 68–69, 78
Nixon, Rob, 33
Nussbaum, Martha C., 113

Ocasio-Cortez, Alexandria, 110, 118,
    175, 179
occupations (climate movement
    tactic), 174–176
Oliver, Mary (poet), 112
Oreskes, Naomi, 76
Overton Window, 110

Pacifism as Pathology (Churchill),
    186
Parekh, Payal, 181
Parenti, Christian, 30, 31
Paris Agreement, 29, 56, 62, 63, 65,
    66, 88, 107, 111, 173
Pelosi, Nancy, 124, 175, 179
People's Climate March, 10, 169–170
Peterson, Jordan, 87
pipelines, 31, 38, 102, 123, 164, 174,

175, 176, 177, 178, 179, 187
planetary boundaries framework,
    135
Political Compass, 24
political spectrum, 22–24
Pollin, Robert, 147
post-truth, 87
Prakash, Varshini, 127
Prevent (UK counter-terrorism pro-
    gram), 187
public banking, 121

race, 4, 16, 18, 22, 30, 195
    climate movement and, 179–184
Raworth, Kate, 142
RCMP, 176, 177, 187
refugees, 31, 32, 86, 88, 163, 171, 182
regulations, 6, 18, 23, 38, 40, 48, 49,
    56–58, 64, 72, 74, 76, 94, 95,
    104, 107, 114, 118, 157, 159
renewable energy portfolio standard,
    57
Republicans (American political par-
    ty and voters), 71, 79, 97, 110,
    111, 123, 126, 127, 175, 191
retrofitting, 122, 123, 160
Reznicek, Jessica, 187
rights, 15, 16, 22, 32–34, 37, 71–73,
    100, 107, 111, 112, 117, 119,
    128, 156, 158, 177, 178, 182,
    183, 185, 187, 190, 192, 193
right-wing ideology
    attacks on Green New Deal based
        in, 123–124
    authoritarian potential in contem-
        porary times of, 86–88
    denial machine and, 75–80
    main beliefs of, 71–74
    media biased in favour of, 68–69,
        82–84, 87, 88, 123–124
    susceptibility to climate change
        denial and, 70–71, 74–75

Rise Up Movement, 179